Automation and Digitization in Timber Construction

Andreas Heinzmann · Niki Karatza

Automation and Digitization in Timber Construction

 Springer

Andreas Heinzmann
Memmingen, Germany

Niki Karatza
Stephanskirchen, Germany

ISBN 978-3-658-47129-3   ISBN 978-3-658-47130-9   (eBook)
https://doi.org/10.1007/978-3-658-47130-9

This book is a translation of the original German edition "Automatisierung und Digitalisierung im Holzbau" by Andreas Heinzmann and Niki P. Karatza, published by Springer Fachmedien Wiesbaden GmbH in 2022. The translation was done with the help of an artificial intelligence machine translation tool. A subsequent human revision was done primarily in terms of content, so that the book will read stylistically differently from a conventional translation. Springer Nature works continuously to further the development of tools for the production of books and on the related technologies to support the authors.

Translation from the German language edition: "Automatisierung und Digitalisierung im Holzbau" by Andreas Heinzmann and Niki Karatza, © Der/die Herausgeber bzw. der/die Autor(en), exklusiv lizenziert an Springer Fachmedien Wiesbaden GmbH, ein Teil von Springer Nature 2022. Published by Springer Fachmedien Wiesbaden. All Rights Reserved.

This Springer imprint is published by the registered company Springer Fachmedien Wiesbaden GmbH, part of Springer Nature.
The registered company address is: Abraham-Lincoln-Str. 46, 65189 Wiesbaden, Germany

If disposing of this product, please recycle the paper.

Preface

Wood construction has steadily gained importance in recent years and is increasingly in the focus of politics. This is becasue the resource saving construction method creates many types of buildings, including living spaces, while also having positive effect on environmental and climate goals. Thus, not only is the demand for wooden buildings growing, but so is their area of application. Consequently, companies specializing in wood construction are facing new challenges in terms of complexity and capacity.

In the context of research, teaching, and consulting, we have gained deep insights into the timber construction industry, numerous corporate structures, and specific problem areas over the past few years. In doing so, we realized that many companies lack knowledge about automation solutions. Thus, they lack the necessary understanding to classify manufacturing concepts, technologies, and processes in terms of their own production requirements. We recognize an even greater deficit in dealing with the topic of digitization because many companies are unaware of the added value that custom software solutions can bring to the value chain. Therefore, this work is intended to provide an overview of possible systems with which the processes of prefabrication and the entire process chain of a construction project can be optimized.

With our joint work in research for timber construction and in the start-up founded from it, our shared vision is clear: We aim to make our contribution to climate-friendly construction by advancing timber construction. Due to the widespread craft structures in timber construction, many companies find it difficult to sensibly expand in terms of personnel and technology. By sharing our knowledge through collected experiences, we want to help gain a broader understanding of automated and digital solutions. Because the introduction and expansion of these areas hold enormous potential for prefabrication in timber construction. However, this can only be fully used to good advantage by considering the entire process chain of construction.

Rosenheim, Germany   Andreas Heinzmann
                                                                     Niki P. Karatza

Contents

Glossary and Terminology in the Context of Timber Construction

BIM:	Building Information Modeling (German: Bauwerksdatenmodellierung, a method for digitally networked building planning)
BTL, BTLx:	Standard exchange format for data in the timber construction sector
CAD:	Computer-Aided-Design (Eng. computer-aided design)
CAM:	Computer-Aided-Manufacturing (Eng. computer-supported manufacturing)
CNC:	Computerized Numerical Control (i.e., computer-aided numerical control or machine control by converting control commands into motion sequences)
Durchlaufzeit:	e.g. the time from the start of production to the loading of the elements
Element:	two-dimensional pre-assembled module of a building
EPS:	Expanded Polystyrol
Fertigungslos:	Summary of parts for joint processing/manufacturing/assembly
Finish:	Completion of e.g. exterior walls
FTF:	Driverless transport vehicle
FTS:	Driverless Transport System
Gefach:	Area encompassed by truss parts, which is usually filled with insulation material (timber frame construction)
Handling:	Movement and handling of components
HOAI:	Fee Schedule for Architects and Engineers (in Germany)

IFC:	Industry Foundation Classes (open standard for data exchange in the construction industry, for BIM models)
KVP:	continuous improvement process
Langteile:	Rod-shaped components with a length of >3.5 m (threshold, frame, etc.)
Los:	see production lot
MES:	Manufacturing-Execution-System (Eng. Manufacturing Management System)
Nest:	Nested cutting plan for e.g. a CNC cutting system (see also nesting system)
OSB3:	Oriented Strand Board, 3: Coarse particle boards for load-bearing purposes for use in damp areas
Referenzierung:	Definition of the position of a workpiece in relation to the coordinate system of a robot or a CNC machine
Taktzeit:	the defined time for delivering the required performance, after which components leave a workstation and are further processed at the downstream location
Verschnitt:	Non-usable material remnants that occur during cutting
Vorfertigung:	Factory production of wooden construction elements
x-Richtung:	along the component
y-Richtung:	transverse to the component
z-Richtung:	perpendicular to the component

Introduction 1

The quota and number of approved residential and non-residential buildings constructed using timber has been steadily increasing in Germany in recent years (Holzbau Deutschland 2022). This trend reflects the growing demand for resource-efficient timber buildings. However, the increasing orders pose a challenge for many companies. Repeated economic surveys by Holzbau Deutschland (2022) have shown that one of the biggest obstacles to success in the industry is the ongoing shortage of skilled workers. As a result, increasing the workforce is not readily possible, and the required capacities to meet new orders is often unachievable. In particular, manufacturing companies dealing with highly crafted structures are reaching their limits. The use of automation is indespansable in order to expand production capacities with flexibility and independently of personnel deployment. However, increasing the degree of automation is only possible if digital processes are used appropriately from planning to data provision to the machines. Many companies therefore perceive the entry into automation and digitization as a hurdle. Primarily due to the lack of digital know-how and a holistic understanding of the technical possibilities. The classification of the various systems in relation to different production requirements and their integration into existing processes is also challenging for many companies.

This document deals with the core processes of timber prefabrication and provides an overview of possible automation and digitization solutions. The focus is on the production of prefabricated stick frame interior and exterior walls, ceilings and roof elements. The processes for factory assembly of the elements into 3D modules are not considered. The existing technologies in the context of the manufacturing processes of complex timber frame construction are examined in more detail. Both the processes and the technologies are occasionally applicable for the

A. Heinzmann and N. Karatza, *Automation and Digitization in Timber Construction*, https://doi.org/10.1007/978-3-658-47130-9_1

extended prefabrication of cross-laminated timber and solid wood elements; such as lumber and heavy timber. Traditional manual processes are only mentioned in passing, as the focus is always on automated tools, machines and systems. The outlined solutions are well-known systems on the market. Including, concepts of special machine construction as well as approaches of research and development.

This work explains how general manufacturing forms and organizations in timber prefabrication are already being implemented and what further visions could be realized in the future. Suitable technologies for all sub-processes of timber frame construction with their respective output (according to VDI 3415, sheet 1) are presented. The output values are based on the long-term experience of the author and relies on different manufacturer specifications from the field of production planning and optimization for timber construction. Finally, the systems are compared with each other in an evaluative manner and their applications are discussed. Since both the presentation and the evaluation and classification of the technologies are also based on the knowledge of the authors, no claim to completeness is made. However, they provide an overarching understanding.

Prefabrication in Timber Construction

2

Wooden buildings, due to their light weight and the good processability of the materials, are particularly suitable for the prefabrication of individual elements. Especially with higher complexities of components and elements, the importance of optimal manufacturing conditions increases. A weather-independent, technology-supported production is only guaranteed in production halls. Wooden buildings are mainly prefabricated in the form of wall, roof, and ceiling elements in panel construction. The load-bearing structures can be both stick-frame components (Fig. 2.1) and solid wood elements (e.g., cross-laminated timber). Both systems are equipped and clad with additional components depending on the degree of prefabrication. (Schankula 2012).

To avoid special transports, the prefabricated elements are usually up to 3.5 m wide and up to 13 m long. The transport of larger dimensions is only possible by obtaining special permits.

The large components are traditionally placed on production tables and assembled into elements in a flat position. The use of aids, machines, and equipment makes the tasks of prefabrication more ergonomic and easier for the workforce than on the construction site. At the same time, better safety practices are achieved. Quality control can also be carried out more reliably with structured prefabrication procedures.

In Fig. 2.2, an example of a conventional assembly table for element production is shown. The illustrations in this work are based on the coordinate system shown. The x-axis runs along the long side of the element. It defines the feed direction of workpieces, components, and elements through machines, as well

A. Heinzmann and N. Karatza, *Automation and Digitization in Timber Construction*, https://doi.org/10.1007/978-3-658-47130-9_2

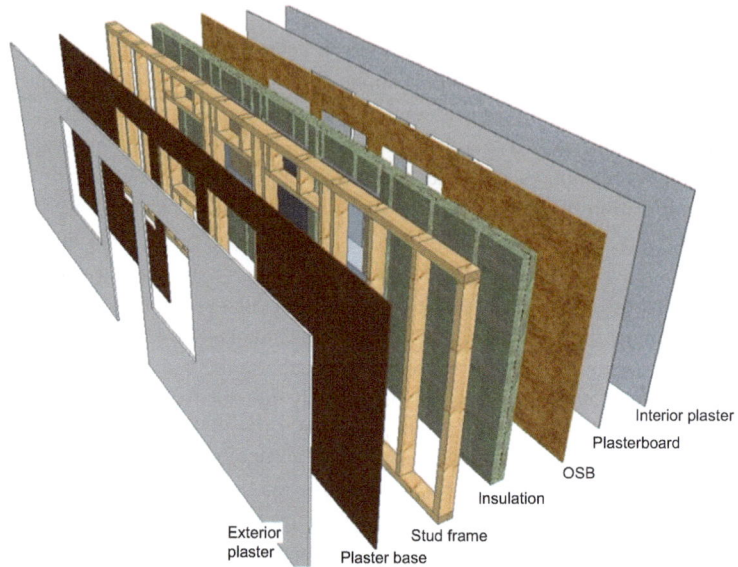

Interior plaster
Plasterboard
OSB
Insulation
Exterior plaster
Stud frame
Plaster base

Fig. 2.1 Exemplary structure of an exterior wall in timber frame construction (Karatza 2019)

as during longitudinal transport in production. Transverse transports or movements accordingly run in the y-direction, while the z-axis describes the vertical orientation.

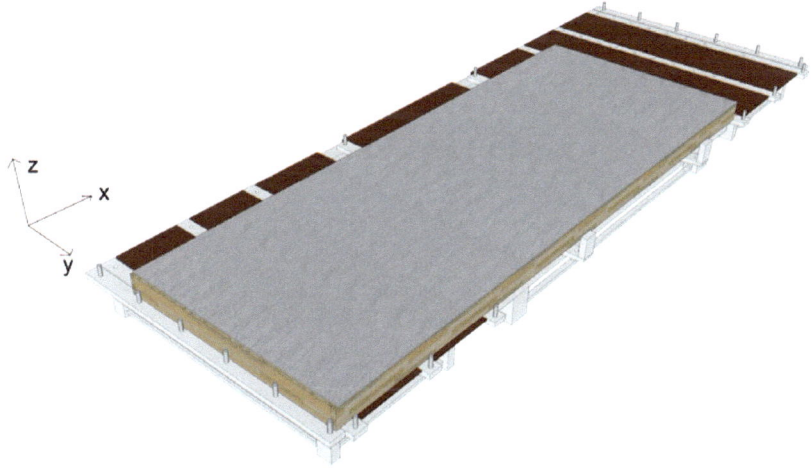

Fig. 2.2   Assembly table for timber construction including coordinate system

Types of Manufacturing

3

The term manufacturing form refers to the combination of manufacturing organization and arrangement of workstations for the processes to be carried out (Eversheim and Schuh 1999, p. 9–66). The selection of a suitable manufacturing principle for a company forms the basis for technology design in production and logistics. It has implications for the possible product variants and manufacturing lead times. In timber construction, manufacturers of highly standardized buildings with low variance have completely different requirements than, for example, a manufacturer of individual construction projects.

The manufacturing forms relevant for prefabrication in timber construction are listed in Fig. 3.1.

3.1   Workshop Production

Workshop production combines work areas and resources for the same processing procedures. Usually, the processing of a station is carried out in batches and collected downstream. Leading to long lead and buffer times for the parts (Dolezalek and Baur 1973, p. 134 ff.). This principle is primarily suitable for the production of small quantities with high individuality and/or complexity. It is less suitable for mass production, as the cycle times can hardly be standardized (Bauernhansl 2020, p. 139).

For implementation in timber construction, it is conceivable to divide individual processes or process groups among several specialized workstations in production, as shown in Fig. 3.2. The individual workstations do not have to be

© The Author(s), under exclusive license to Springer Fachmedien Wiesbaden GmbH, part of Springer Nature 2025
A. Heinzmann and N. Karatza, *Automation and Digitization in Timber Construction*, https://doi.org/10.1007/978-3-658-47130-9_3

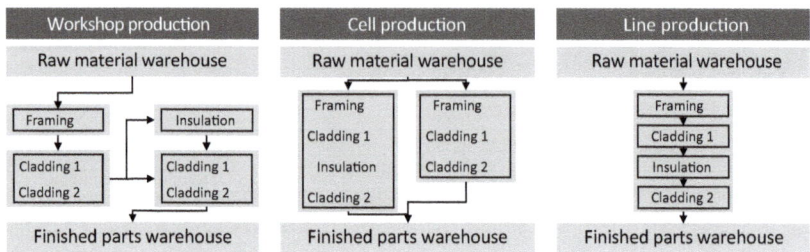

Fig. 3.1 Manufacturing forms for prefabrication in timber construction. (Based on Bauernhansl 2020, p. 139)

rigidly linked, which makes the process sequence flexible. However, batch production in timber construction is not feasible, as the space requirement of the buffer areas of each workshop would be enormous due to the large dimensions of the elements. Nevertheless, the utilization of the stations can be ensured with individual buffers.

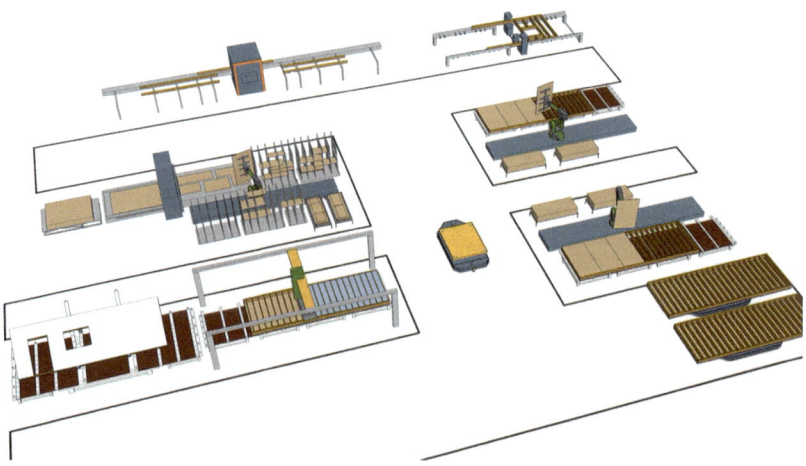

Fig. 3.2 Example of workshop production in timber construction

3.2   Cell Production

Island, group or cell production refers to the spatial bundling of machines and workstations for the production of a product (Bauernhansl 2020, p. 139).

In timber construction, this manufacturing principle can be implemented with varying degrees of automation in both craft and industrial structures. The elements are completed in several processing steps using different machines at a single workstation (example Fig. 3.3). All required materials and components are therefore only to be provided at one place of production. The production cells are also decoupled from upstream and downstream production processes, which particularly includes logistics, material handling, and finishing. To ensure that the cell is supplied with the necessary components in time, regardless of the status of these areas, and that the elements are immediately removed after completion, the use of buffers for the cut parts, raw materials and finished parts is necessary.

The individual processes that can be a variety of influencing factors such as:

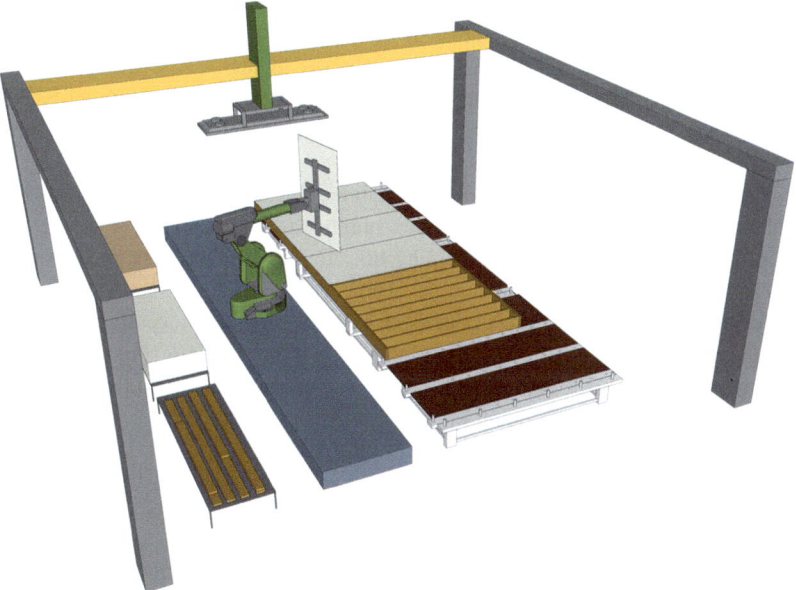

Fig. 3.3   Example of cell production by robots in timber construction

- the degree of prefabrication of the elements,
- the execution of individual element layers (e.g., plaster or wood facade),
- the degree of automation of the cell processes,
- the machines used and their technical possibilities.

3.3   Line Production

In line or flow production, the individual process areas are spatially arranged in the order of the production flow, which is usually implemented in a straight line, U-shape, or circular form. Furthermore, the respective processing times should be coordinated as closely as possible with the cycle time. This coordination represents the greatest challenge of line production, because as soon as the process times of the stations differ from each other, waiting times arise for individual areas. (Dolezalek and Baur 1973, p. 138 ff.)

This form of production is so far most widespread in the industrial prefabrication of wooden building elements. The assembly processes are divided into individual successive production tables or areas and are dependent on each other in the production flow. The elements are transferred directly from station to station using integrated conveyor technology in the tables (e.g., chain conveyors, driven rollers, or belts) and are only buffered in the rarest of cases. Which makes the provision of the required materials at several points across the assembly process crucial and necessary.

The breakdown of work content into stations can be detailed or combine several activities, which can also vary the number of tables or work areas. The decisive factor is the average processing time for the respective work content to be carried out at the individual stations. They should not differ significantly from each other for as smooth a cycle as possible. The longest processing time of a single area is always the pacemaker of the entire process chain (Fig. 3.4).

Line production can be found in timber construction in two different forms: in the form of several individual tables or as a single long table.

3.3.1   Line through Individual Tables

A common structure of line production with several linked individual tables is shown in Fig. 3.5. The work content is assigned to a table according to the required performance and the degree of prefabrication. Due to the maximum

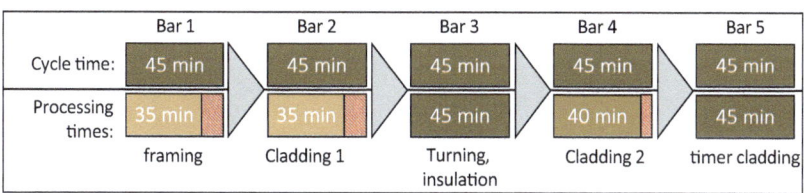

Fig. 3.4 Cycle timing in timber prefabrication: processing and waiting time in relation to cycle time

lengths of elements produced, the individual tables usually have a fixed length of about 12 m.

3.3.2 Endless Table

In cycle production on a single long table, the elements pass through specialized areas of the table to which the broken down processes are assigned. The table consists of several combined segments and can thus be created in unlimited length. The length of the sub-areas is defined based on the required performance or the corresponding cycle time. However, due to the uninterrupted work surface, one is not bound to a fixed table grid (see Sect. 3.3.1), so that the flexible production of elements of different lengths is possible. The sub-areas are each equipped with various technical components such as transport units for the longitudinal and transverse conveyance of the elements or turning systems. An endless table also has a multitude of possible zero points, so that the elements can be referenced in each area (Fig. 3.6). This is essential for the use of automation solutions.

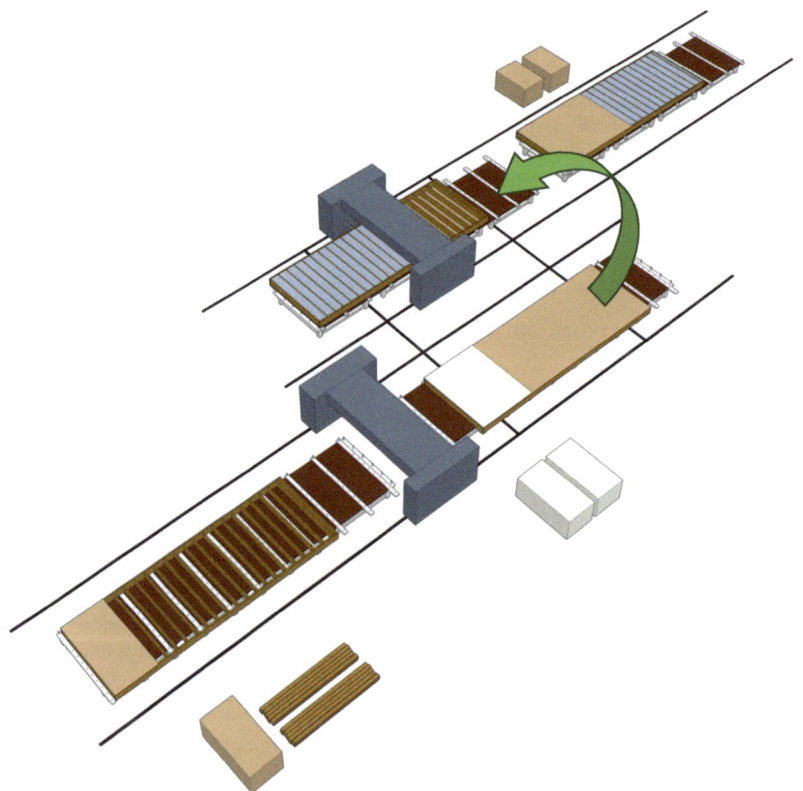

Fig. 3.5 Example of line production in timber construction with individual tables

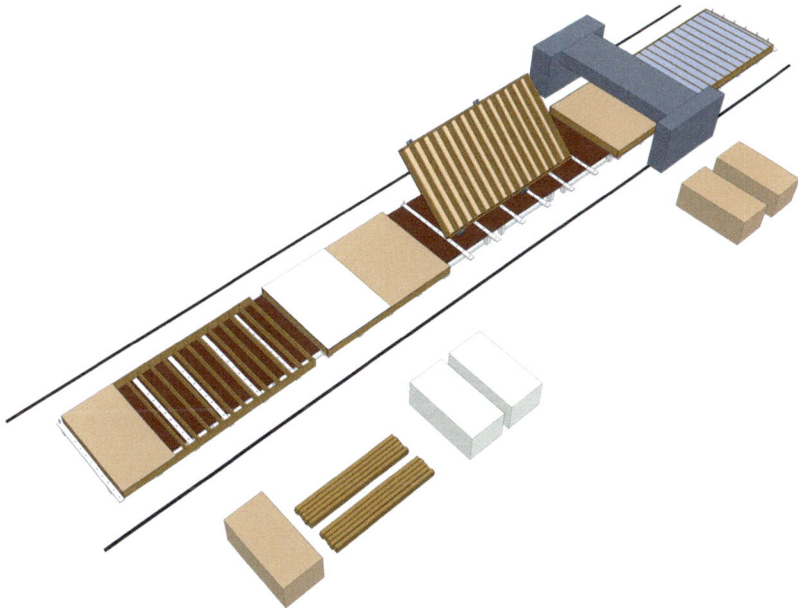

Fig. 3.6 Example of line production in timber construction with endless table

Processes and Manufacturing Technologies

4

This chapter describes the main processes of a timber frame prefabrication as well as suitable technologies for it. The most common levels and systems of automation are outlined and compared and evaluated based on relevant criteria. The essential material forms in timber construction are wooden beams, studs or panels. Due to this, their processing and assembly processes can occasionally differ from each other, which is why they are sometimes listed separately. Figure 4.1 shows the process flow considered in the following, which depending on the manufacturing organization, begins with the supply of raw materials to the cutting plant or with the provision of materials at the stations. The options for facade design conclude the process.

4.1   Material Storage and Supply

The type of supply of raw materials to a plant depends on several factors. On the one hand, the storage and supply beams, studs and panels can differ. On the other hand, the process depends on the machine's mode of operation. For example, panels can be cut both in stacks and individually. Furthermore, the process is influenced by whether the material procurement is always in the same standard dimensions or project-based, e.g., in several defined lengths. Materials can therefore either be stored together and taken directly for processing or require a sorted supply. For the latter, there is the possibility that the individually prepared materials are already delivered commissioned and prepared for automated removal as needed. If this is not guaranteed by the supplier, the sorting according to the processing sequence for machine feeding must take place in the factory. It must be

A. Heinzmann and N. Karatza, *Automation and Digitization in Timber Construction*, https://doi.org/10.1007/978-3-658-47130-9_4

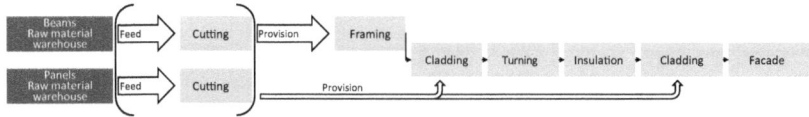

Fig. 4.1   Core processes of a timber frame prefabrication

weighed whether the supply is more sensibly automated or manually designed. If sorting and provision are very complex or time-consuming, an automated solution should be considered.

4.1.1   Storage and Supply of beams

In timber frame construction, solid timber was the most common pole-shaped material for a long time. However, in recent years, developments in timber construction towards more complex and taller buildings have been observed. The number of multi-storey residential buildings in timber construction has tripled since 2011, for example (Federal Statistical Office 2020). This also changes the requirements for the components. At the same time, changes and fluctuations in the raw material market have effects on material availability and quality (Jakob 2021), which is why new materials are increasingly being used in timber construction in great variance. With the increasing fluctuation of raw materials and the complexity of the components, the relevance of process automation increases. Project-independent procured standard materials or procured parts with a project reference, which were not cut exactly by the supplier, are cut and processed in production in advance for the assembly of the elements. The type of provision of the beams and studs, not least depends on the variance of the row material. For cutting optimization, a larger selection of different raw materials is useful. Therefore, a material is often procured in various lengths and dimensions. When cutting different studs and beams, it must be ensured that the right dimension is delivered to the machine at the right time. There are different options for provision and supply.

4.1.1.1   Manual Rod Supply using Side Loaders

The supply of rod-shaped components by a side loader is the most widely used method (see Fig. 4.2). Because it is carried out manually by employees, the

Fig. 4.2 Rod supply using side loaders

provision is very flexible. However, especially for the timely supply of individual dimensions from a large variance, a much higher personnel deployment is required than with automated solutions. Also, the commissioning before cutting is only possible manually with this type of supply.

4.1.1.2   Rod Feeding through Automatic Area Portal with Individual Rod Handling

In the variant of individual rod handling, an area portal moves the rods into an intermediate storage or into the feed area of the machine. The delivered material stacks are placed on a storage area (Fig. 4.3, left), from where the portal picks up each rod individually. The sorting of the individual rods is done according to dimension and length in defined rack compartments. From there, the required individual rods are removed and fed to the machine.

A rack area is usually pure in type. However, for rarely used cross-sections, it is also possible to form mixed stacks in one compartment and to sort them, for example, overnight. The sorting of the rods from the storage area into the racks can take place while the machine is processing a piece of wood. Because then it has no direct need for material. If several storage areas are available, it is possible to place stack packages on them, from which the individual rods can be directly fed to the machine. The saving of the previous storage in racks reduces the process time and increases the output.

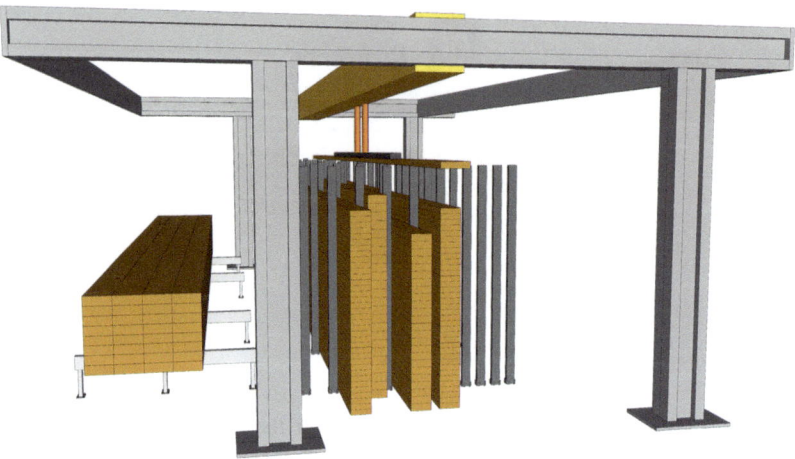

Fig. 4.3 Rod feeding through area portal with individual rod handling

Application (according to VDI 3415, sheet 1) per shift (8 h):	250-300 loading processes of individual bars
Number of employees:	0 (only for monitoring and providing new material stacks, as the process is automated)
Note on application:	• Specification for one storage transfer area • Each beam must be picked up 2 times • the output depends on the size of the gantry and the height of the stacks in the stanchions, as this has an influence on the distances to be covered

4.1.1.3   Rod Feeding through Automatic area Portal with Removal of Entire Stack Layers

In the case of feeding the rods through an area portal with layer-by-layer removal, the components are stored in stacks. A suction device removes an entire layer of wood and places it in the feed area of the machine (Fig. 4.4). The advantage here is that the separation of the rods is eliminated, thereby reducing handling. However, due to the increased space requirement for stack storage under the portal, only a limited variance of rods can be provided. The system is therefore primarily

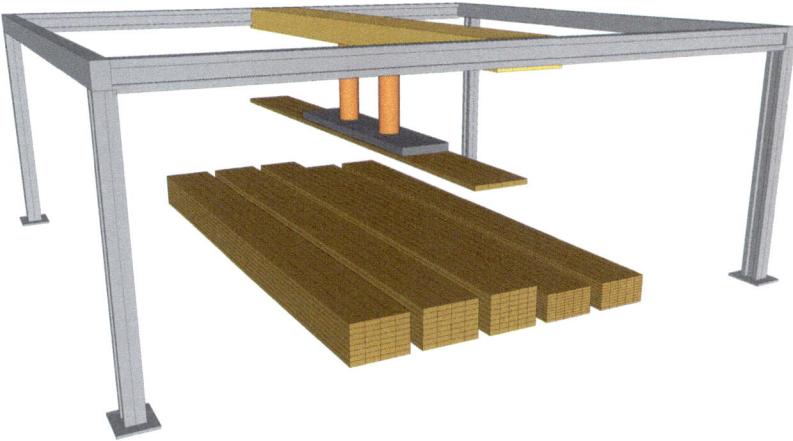

Fig. 4.4 Rod feeding through area portal with layer-by-layer removal

suitable for the main use of standard dimensions. A risk of layer-by-layer handling is that particularly twisted rods can detach from the vacuum suction device.
This reduces process reliability. It is also important to note not to make the area
storage too large, so that the travel times of the portal can maintain the necessary cycle performance. With appropriate design, the portal can also be used for
feeding several cutting machines.

Application (according to VDI 3415, sheet 1) per shift (8 h)	250-300 loading processes of layers with several bars
Number of employees:	0 (only for monitoring and providing new material stacks, as the process is automated)
Note on application:	• the output depends on the size of the portal, as this has an influence on the distances to be covered

4.1.1.4   Comparison and Classification of Rod Feeding Systems

A side loader is a common entry point for in-house material logistics. Even
after an investment in automatic feeding equipment, it is useful to use it, for

Table 4.1 Comparison of rod feeding systems (evaluation of criteria from excellent (+++) to poor (− −))

Criteria	Side loader	Area portal single rod	Area portal layer-wise
Flexibility	+++	++	−
Staff deployment	−	++	++
Level of automation	− −	+++	++
Investment	++	−	−
Future viability	−	++	+
Space requirement	+	−	−
Data requirements	+++	−	−
Process reliability	+++	+	−

example, to move stacks to a storage location. However, the significantly growing need for variance in raw materials increases the requirements for logistics in the machine environment and thus significantly the effort of manual feeding. Therefore, automation solutions for material feeding are not only useful for larger companies, but generally (Table 4.1).

When using area storage systems, it should be noted that their capacities are limited due to space constraints. Thus, they only form a buffer from which the plant can be supplied with material for a few work shifts and do not house the entire raw material warehouse.

Furthermore, it should be taken into account that automated technologies require increased demands on warehouse management and cutting data. In particular, a cut-optimized cut requires different lengths of a material, which must accordingly be managed in the buffer.

4.1.2 Storage and Feeding of Plates

There are several factors to consider when feeding plate materials into the processing machines. In addition to the dimensions and plate thickness, the material density and strength must be taken into account for handling. It is necessary to distinguish which materials can be suctioned and which, on the other hand, must be gripped, such as certain air-permeable softwood fiberboards. The type of processing in the machine determines whether the feed is as a stack or a single plate.

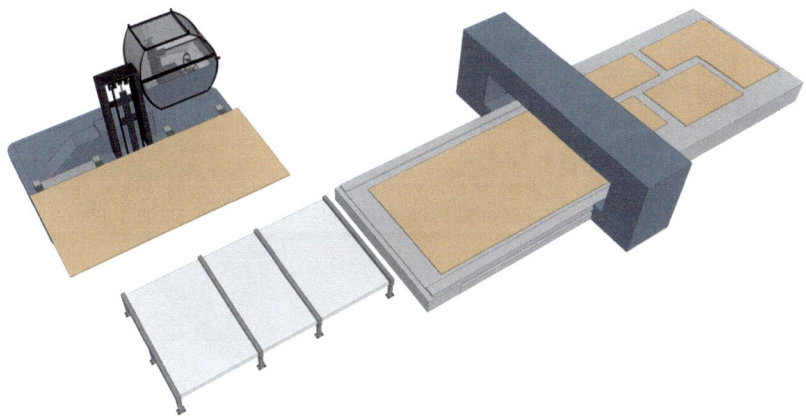

Fig. 4.5 Plate feeding using a forklift

4.1.2.1 Manual Plate Feeding Using a Forklift from Block Storage

Plate feeding from a block storage by employees using a forklift is so far the most widespread method, as it is very simple and flexible (Fig. 4.5). However, the increasing variance in installed plate types and formats increases the need for plate changes at the machine and thus also the necessary personnel deployment. This quickly leads to bottlenecks and affects the performance of the machine.

4.1.2.2 Panel Feeding by Automatic Area Portal

An area storage with feeding to a connected machine runs fully automated (Fig. 4.6). For storage, the delivered plate stacks are placed on a defined storage space, from which an area portal individually removes and sorts the plates. The sorting in the area storage is either pure or in the form of mixed stacks. To avoid re-sorting times by the area portal, pure stacks are usually used. Mixed stacks are useful for rarely used materials in low stock or for plate remnants. Before feeding the mixed plates into the cutting, a re-sorting in the order of processing is necessary. This can be done, for example, at night to avoid waiting times of the machine.

Higher flexibility and significantly higher performance in the feeding process is achieved by using multiple storage spaces or a conveyor line with parking spaces (Fig. 4.6, right). From there, the portal transports the individual plates

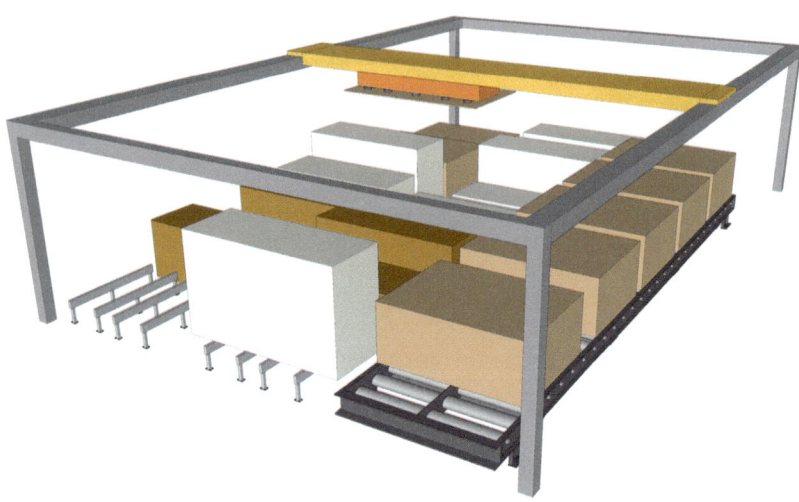

Fig. 4.6 Plate feeding by area portal

directly from the stack to the machine without storage. This approach pays off particularly when several cutting machines are supplied by one area portal.

Application (according to VDI 3415, sheet 1) per shift (8 h):	200-250 loading processes of individual panels
Number of employees:	0 (only for monitoring and providing new material stacks, as the process is automated)
Note on application:	• Specification for one storage transfer area • Each panel must be recorded twice • the output depends on the size of the portal, as this has an influence on the distances to be covered

4.1.2.3   Panel Feeding from High-Bay Warehouse

If there is the possibility to take advantage of a hall with enough height, a high-bay warehouse can be used for plate storage and feeding, reducing the space requirement in the area (Fig. 4.7). The shelving systems are usually standard components, where the compartments can be individually adjusted in height. This makes it possible to store entire stacks of plates at different heights as well as individual plates. Due to the high number of possible storage spaces, it is usually not necessary to mix different plates on individual stacks. The stacks are therefore usually stored purely by type.

The core of a high-bay warehouse is an automatic storage and retrieval machine. This is equipped with a vacuum lifter and/or a fork for material movement. The vacuum suction device removes the plates individually from the warehouse and places them on a transfer station, from where they are conveyed to the machine table. Alternatively, the system lifts the plates in stacks using a fork and prepares them in this way. It should be noted that stacks whose plates have not been completely used up must be stored again. Depending on the requirements, a combination of handling systems can be used to store plates in stacks

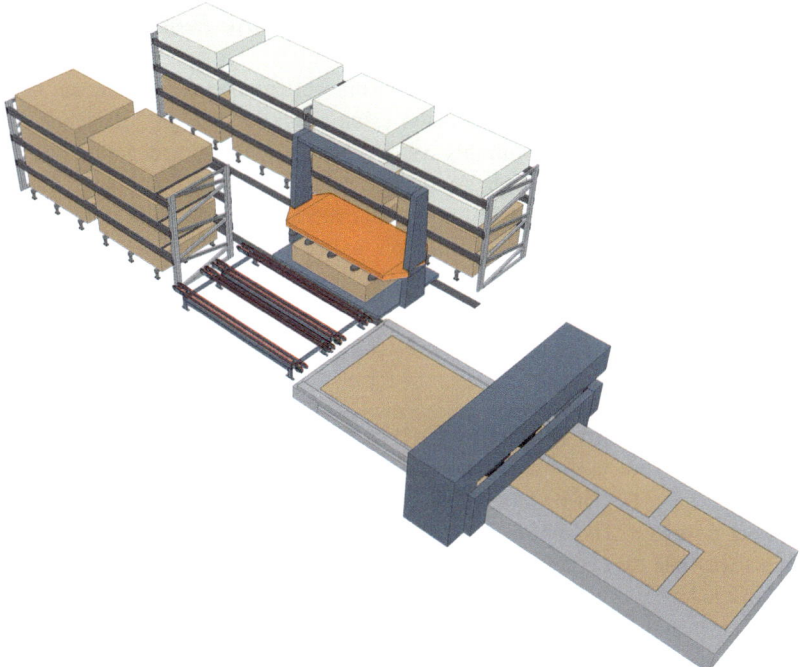

Fig. 4.7   Plate Feeding from High-Bay Warehouse

using a fork, for example, and to remove them individually using a vacuum suction device.

Application (according to VDI 3415, sheet 1) per shift (8 h):	250-300 loading processes of individual panels
Number of employees:	0 (only for monitoring and providing new material stacks, as the process is automated)
Note on application:	• Stack-wise storage and removal of individual panels by the storage and retrieval unit • the output depends on the size of the rack, as this has an influence on the distances to be covered

4.1.2.4 Comparison and Classification of Plate Feeding Systems

The use of many different materials and formats often necessitates frequent plate changes. In particular, manual provision using a forklift requires significant personnel effort. Furthermore, with this system, the risk of machine downtime due to stack change times is very high. Automated solutions, on the other hand, ensure that a machine is continuously supplied with material. Area storage systems are in most cases a sensible automation solution, as they offer sufficient performance, capacity, and flexibility for a medium investment sum. In the case of a high variance of raw plates and/or limited space conditions, the high-bay warehouse proves to be a suitable alternative (Table 4.2).

4.2   Upstream Material Cutting

Individual processes of prefabrication in timber construction can be usefully carried out upstream. The aim is to prepare materials, components or assemblies in such a way that they relieve the assembly processes and thus shorten the throughput times. Advantages can be achieved in timber construction specifically through upstream material cutting. Since most building elements differ from each other, materials must in any case be individually cut and processed for their respective installation. If these cutting operations are separated from the element assembly, the waste and residue handling of materials is eliminated there. Therefore, dust

Table 4.2 Comparison of plate feeding systems (evaluation of criteria from excellent (+++) to poor (– – –))

Criteria	Forklift	Area Portal	High-Bay Warehouse
Flexibility	+++	++	++
Variability for many different materials	–	++	+++
Investment	++	+	–
Future Viability	–	++	++
Space Requirement	+	–	++
Data Requirements	+++	–	–
Process Reliability	+++	+	+

and chip extraction is only necessary at the cutting work area, but not in further production areas.

Another significant benefit is the possibility of optimized material cutting. In order to make optimal use of the material through a cutting system, the parts to be cut are planned digitally in advance in a CAD/CAM system and combined into production lots (e.g., all panels of a material for six walls). Another software then processes the specifications of all parts of a lot and creates a cutting image that optimally utilizes the raw material. In this way, several components and as little waste as possible are produced from one piece of raw material.

The optimized material cutting in production lots results in the parts not being processed in the order in which they are installed in the prefabrication. The individually processed components must therefore be picked, sorted and made available correctly and on time for the respective workstation. Overall, it can be said that the larger the production lot or the number of individual parts to be cut together, the better the waste optimization, but the higher the subsequent sorting effort.

The choice of technology, as with most processes, depends on the required performance, the respective materials, and the complexity of the processing. These can be simple right-angled cuts, angled cuts, notches, drillings or more specific processing such as milling cable channels in an insulation board of the installation level. Depending on the detail of this process, the components are reprocessed in assembly or simply installed.

4.2.1   Rod Cutting

The cutting of rod-shaped components is only considered upstream, as timber frame prefabrications are almost exclusively organized in this way. The rods are combined into lots optimized for waste and cut automatically. Since the large dimensions are difficult to handle for subsequent sorting, the cutting of the long parts usually does not take place in lots with small parts. Instead, they are cut in the required assembly order. The procurement of different raw material lengths reduces the waste by only producing short remnants with the appropriate choice of input length.

4.2.1.1   Cutting System for Rods

The lowest level of automation for the cutting of rod-shaped components is the cutting system. In addition to manually operated machines, there are semi-automatic saws. They support the positioning of the components and cut the woods automatically with data connection and electronic stop. The data transfer also allows for waste-optimized cutting through the processing of cutting lists. These saws are suitable for simple cross and miter cuts. Depending on the design, several woods can be cut at the same time when stacked on top of each other (Fig. 4.8).

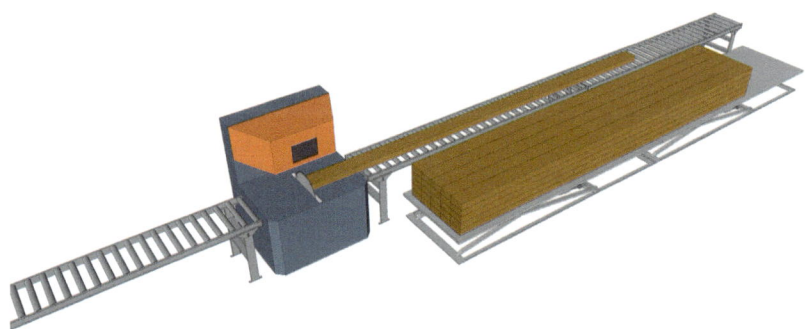

Fig. 4.8   Rod cutting by cutting system

Application (according to VDI 3415, sheet 1) per shift (8 h):	up to 1,000 simple cuts
Number of employees:	2 (for maximum power)
Note on application:	• Capacity is determined by the operators, who usually also perform logistics and sorting tasks such as manual parts supply and removal

4.2.1.2   Rod Cutting System with Extended Processing

If further processing is required beyond simple length cutting, the use of a fully automatic cutting system is beneficial (Fig. 4.9). In addition to arbitrary angle and inclination cuts, this system also carries out simple drilling and milling operations. Machines of this type ensure high precision and performance in positioning, handling, and cutting. The processing of several stacked woods is also possible here. Conveyor systems take over the feeding of the workpieces as well as the removal of material residues, chips, and dust. With the use of such a machine, a high throughput is possible for simple components such as posts, sills, and frames for timber frame construction.

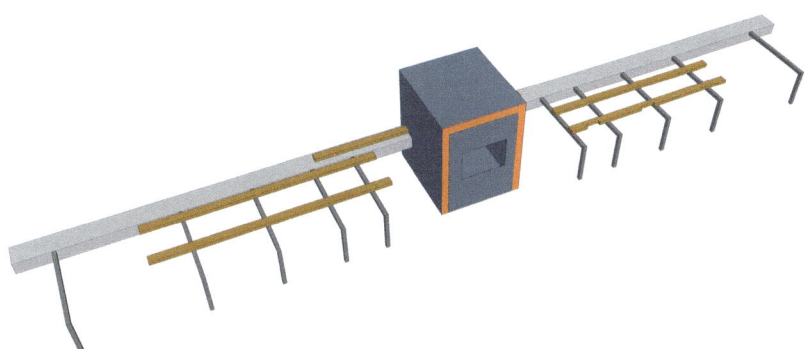

Fig. 4.9   Rod cutting by cutting system with extended processing

Application (according to VDI 3415, sheet 1) per shift (8 h):	Up to 1,000 simple cuts
Number of employees:	1
Note on application:	• The capacity-determining factor here is the complexity and frequency of the extended machining operations such as drilling or milling • Automatic feeding of bars via lifting table or gantry

4.2.1.3   Carpentry Machine

The highest flexibility for rod cutting is offered by a carpentry machine (Fig. 4.10). There are different machine construction concepts. On the one hand, different processing units can be connected in series, with the workpiece being conveyed through the plant from one processing area to the next. On the other hand, there are concepts based on a tool change system, where the workpiece processing is limited to a single area. Different tools such as saw blade, milling cutter, drill, etc. are clamped in a machining spindle. Feeding, positioning, cutting and processing as well as removal are fully automated. The plants offer versatile processing options for complex components.

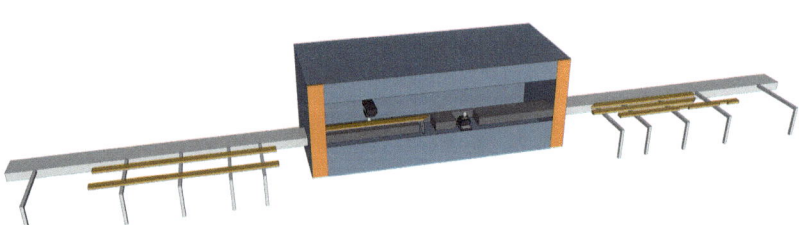

Fig. 4.10   Rod cutting by carpentry machine

Application (according to VDI 3415, sheet 1) per shift (8 h):	Up to 500 blanks with more complex processing
Number of employees:	1
Note on application:	• The capacity-determining factor here is the complexity of the machining operations • Complex components can take several minutes to process • Automatic feeding of bars via lifting table or gantry

4.2.1.4   Comparison and Classification of Rod Cutting Systems

Depending on the requirements for performance and processing, the combination of several machine concepts in one production is possible. For example, a joinery machine can primarily be used for the processing of components for roof and ceiling elements, while a cutting machine takes over the wall production. More complex components of wall production can still be created by the joinery machine. However, with the simultaneous use of several machine concepts, the requirements for data generation increase. During the planning of a building, the feature is set in the CAD/CAM system on which machine the components should be manufactured. So far, the assignment is done manually, as no available software exists. Ideally, a flexible automatic assignment based on machine availability and component requirements would be ideal (Table 4.3).

Table 4.3 Comparison of different rod cutting systems (evaluation of criteria from excellent (+++) to poor (−−−))

Criteria	Cutting machine	Cutting system	Joinery machine
Space requirement	+	−	−
Waste handling	−	++	++
Output of simple cuts	+	++	−
Processing of complex components	−−	−	+++
Complexity of data connection	+	−	−
Cutting optimization	−	++	++
Compatibility with automation	−	++	++

4.2.2 Panel Cutting

The cutting of panel-shaped materials and components is relevant for both timber frame construction and element production with cross-laminated timber or solid wood parts. The process is implemented both upstream and integrated into pre-assembly. The following section compares the two concepts and shows the use of machines available on the market for upstream cutting.

4.2.2.1 Comparison of Methods for Panel Cutting

The panel cutting on the assembly table is carried out after the paneling of the element. The chosen technology cuts the panels on site, creates cutouts for windows and doors as well as recesses for sockets or similar. This concept does not require any sorting activities or spaces (Fig. 4.11). However, the processing times can disrupt the production flow of a line production as soon as the individual element designs differ greatly from each other. Furthermore, dust, chips, and residues are produced in the manufacturing process, which must be transported or vacuumed away. This is technically difficult to achieve with movable automations and leads to increased dust exposure in the work area.

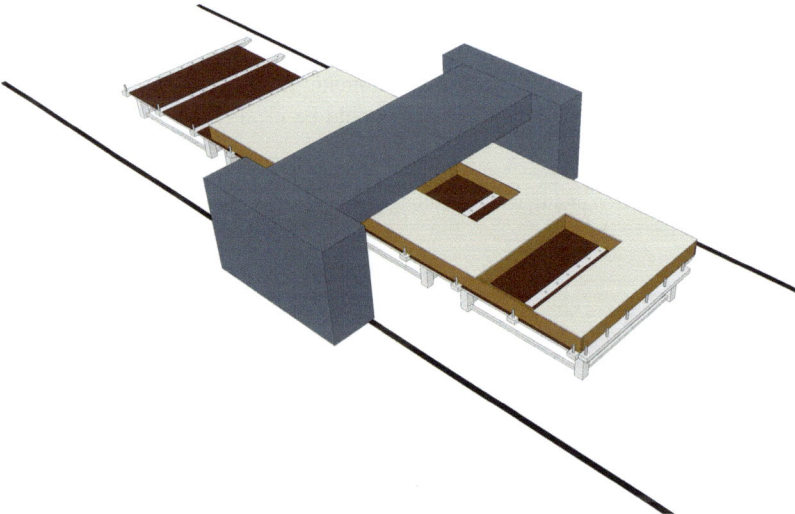

Fig. 4.11 Example of panel cutting on the element table by portal

For the upstream panel cutting, the relevant processing information of all individual panels must be transferred to the system earlier in the process (Fig. 4.12). Therefore, this results in increased work effort for the work preparation. Since the machining of all parts takes place collectively at one location, waste disposal is easier to manage. After the cut-optimized, batch-wise cutting (see Sect. 4.2), the parts are sorted in an additionally required area. The accuracy of the sorting depends on the degree of automation in the environment of the cutting system as well as the requirements by the element assembly (see Sect. 4.3). The sorted parts are then buffered and delivered to the pre-assembly in a timely manner.

The differences between the two variants of plate cutting are summarized in Table 4.4. Both methods were considered in an automated context.

In order to make an informed decision between the methods for a specific production, it is necessary to consider the entire process chain. This is because the individual factors have effects on different areas. While cutting at the element table is less complex in terms of work preparation and logistics, the manufacturing processes are more time-consuming and not optimally calculable. Furthermore, automated laying of uncut plates is hardly feasible due to high tolerances of the input materials. On the other hand, upstream plate cutting requires increased investments and a higher space requirement in the upstream processes. However, the element production and the production flow run more smoothly.

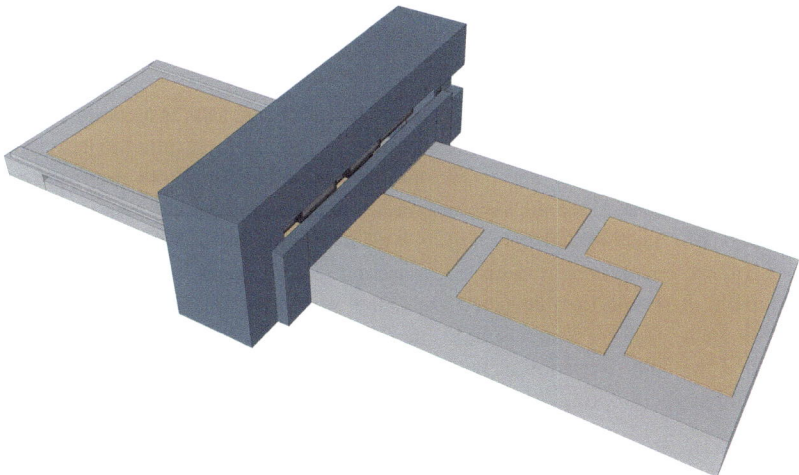

Fig. 4.12 Example of upstream panel cutting on cutting system

Table 4.4 Comparison of different plate cutting methods (evaluation of criteria from excellent (+++) to poor (−−))

Criteria	On the element table	Upstream
Sorting effort	++	−
Space requirement	++	−
Waste handling	−	++
Complexity of material provision	+	−
Complexity of data connection	+	−
Resource utilization	−	++
Compatibility with manufacturing principles	+	+
Compatibility with automation	+	++
Process reliability in automation	−	+

Furthermore, the degree of automation of the production should be considered in the decision-making process. For example, if fully automated work is carried out with robots, it may make sense to outsource machining from this area. Cut plates, which are sorted in the correct assembly order, can be installed fully automatically and reliably in a predictable process.

In summary, it can be said that resource efficiency in the area of raw material clearly speaks for an upstream plate cutting, which also favors downstream automations.

4.2.2.2 Horizontal Panel Saw

The horizontal panel saw, also known as a pressure beam saw, is a compact solution for panel cutting (Fig. 4.13). It delivers precise cutting results for individual panels and package cuts. The panel saw can be loaded from both the front and the back. However, since the saw only moves linearly, the cutting plan is limited, which is why nesting and notching as well as milling or drilling are not possible. If these operations are required, a second cutting process on the element is inevitable. It should then be noted that panel remnants and waste occur at both cutting areas. There is also a double energy requirement for both the cutting process and the chip extraction at both locations. The two-stage process allows for a preliminary, cut-optimized panel cutting despite operations on the element.

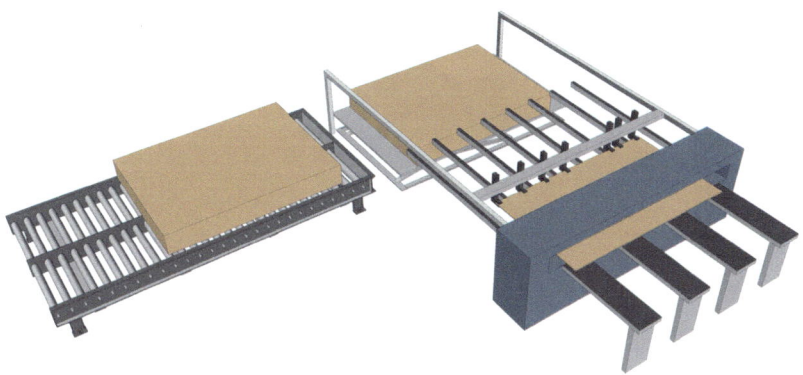

Fig. 4.13 Panel cutting by horizontal panel saw

Application (according to VDI 3415, sheet 1) per shift (8 h):	up to 1,000 panels
Number of employees:	1
Note on application:	• Simple pressure beam saw (no angular saw) • Cutting of individual panels (no book cutting) • Automatic feeding of panel materials via lifting table or automatic storage system

4.2.2.3 CNC Machining Center with Vacuum Table

A CNC machine processes CAD files using control technology and fully automates the machining of workpieces (Fig. 4.14). The panels are placed on the vacuum table of the system, sucked in there and positioned stationary, while the tool units move for machining. It must be ensured that the panels to be processed have a suitable density to be sucked in. The systems can move several tools with up to five axes. This allows for very precise complex cuts, milling and drilling. The machining and formatting of the panels usually takes place using finger milling cutters. If panels of greater thickness such as 80 mm thick softwood fiber panels need to be formatted, the increased risk of fire must be taken into account. Because the larger chip and dust accumulation leads to clogging of the milling paths. Therefore, these cuts should necessarily be taken over by a saw blade. The subsequent removal of the machined panel parts can be done manually or, for example, automated with the help of robots.

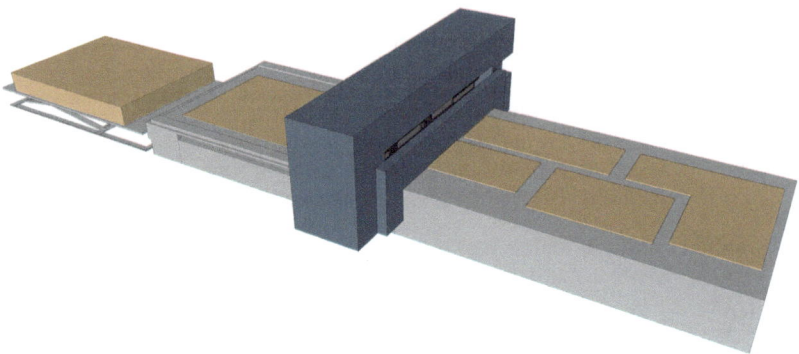

Fig. 4.14 Panel cutting by CNC machining center

Application (according to VDI 3415, sheet 1) per shift (8 h):	up to 500 panels
Number of employees:	1
Note on application:	• Milling for sockets, cable ducts etc. are included in the service on a pro rata basis • Automatic feeding of panel materials via lifting table or automatic storage system

4.2.2.4 CNC Panel Cutting System

In the cutting system described here, the workpiece is held on the side and/or from the back with jaw grippers. The feed of the workpiece takes place in the x-direction, while the machining units for cuts, milling and drilling move in the y- and z-direction. While the use of milling cutters is common in such systems, equipping with tool changers for different units is possible depending on the manufacturer. The combination of movement of workpieces and tools allows the machining of all workpiece shapes in all directions. Thicker panels must also be cut here using a saw blade (see Sect. 4.2.2.3). The cut components can either be removed individually or pushed off as a whole nest. For automated removal

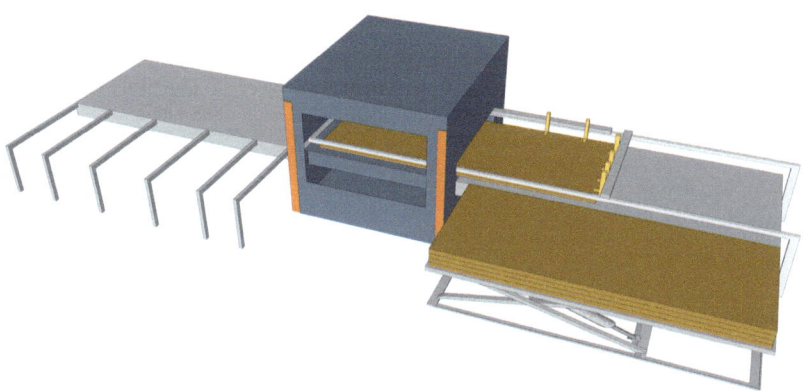

Fig. 4.15 Panel cutting by CNC panel cutting system

by e.g. robots, a separation and generally a referencing of the panel parts is necessary, by driving them against a stop for example (Fig. 4.15).

Application (according to VDI 3415, sheet 1) per shift (8 h):	up to 600 panels
Number of employees:	1
Note on application:	• Milling for sockets, cable ducts etc. are included in the service on a pro rata basis • Automatic feeding of panel materials via lifting table or automatic storage system

4.2.2.5 Comparison and Classification of Panel Cutting Systems

When using a horizontal panel saw for a rough pre-cut, the focus is on optimal material utilization. Other resource-saving criteria such as energy consumption and suction power due to double chip removal are not sufficiently considered. The needs of a resource-optimized production can thus only be partially met. CNC machining centers require that the plates to be processed are airtight and can be suctioned. However, this does not apply to all materials used in timber construction. Holding the plates with grippers as in CNC panel cutting systems

Table 4.5 Comparison of different panel cutting systems (evaluation of criteria from excellent (+++) to poor (− −))

Criteria	Panel Saw	CNC Machining Center	CNC Panel Cutting System
Sorting Effort	++	++	++
Space Requirement	++	+	+
Waste Handling	+	−	+
Chip Extraction	++	−	+
Complexity of Data Connection	++	+	+
Resource Utilization	+	++	++
Compatibility with Automation	+	+	+++
Future Viability	−	+	++

does not limit the machining possibilities due to the materiality of the workpieces (Table 4.5).

4.3   Sorting, Buffering, and Commissioning

The upstream material cutting is carried out as described in cutting-optimized batches with components of several elements. Therefore, a subsequent element-wise commissioning of the cut parts is necessary for a smooth assembly process. Ideally, the parts are sorted in assembly order. The solution approaches shown below are applicable to both bars and panels. Long parts such as ceiling beams and rafters are usually commissioned separately from short parts.

4.3.1   Manual Acceptance with Digital Support

Digital supports are low-automation solutions that indicate to employees, for example, where to sort each component by scanning labels or using light signals (see also Sect. 4.14.4). A prerequisite for this is the development of load carriers that allow each individual part to be sorted into a defined position. The load carriers are precisely planned in advance so that the parts for the assembly sequence are to be placed in predetermined compartments. As soon as a load

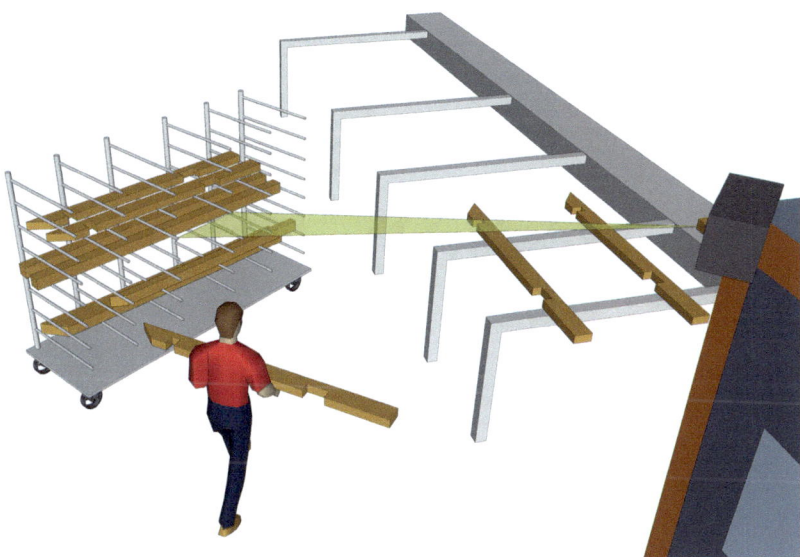

Fig. 4.16 Manual picking with digital support using the example of rods (analogous for plates)

carrier for the respective element has been completely filled, the internal logistics receives a signal and transports the components into a buffer or directly to the element assembly (Fig. 4.16).

4.3.2   Acceptance by Automated Area Portal

An automated retrieval solution is the use of an area portal that removes the components from the machine's output area using grippers or vacuum suckers and sorts them into a buffer (Fig. 4.17). The parts can be placed individually or combined into stacks. The stacks are either sorted by type or, if necessary for optimal space utilization, formed chaotically and resorted as needed. The automatic picking by the portal can stack parts sorted by elements or place them on a conveyor system for connection to the downstream process.

This technology solution is suitable for both long and short components. However, for plate materials, it is only partially useful as the area requirement here is very large.

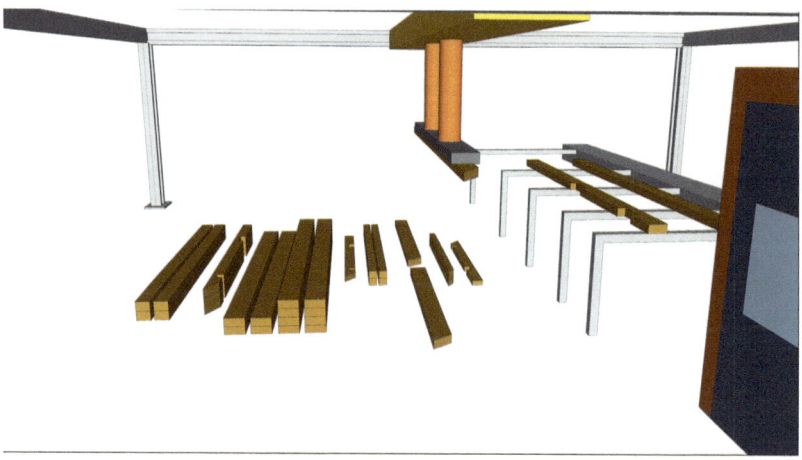

Fig. 4.17 Acceptance by automated area portal using the example of rods (analogous for plates)

Application (according to VDI 3415, sheet 1) per shift (8 h):	400-500 parts
Number of employees:	0 (only for monitoring, as automated process)
Note on application:	• If re-sorting processes become necessary, the performance of the system is reduced • Shorter parts can be moved faster due to lower inertia and therefore increase the possible output • the output depends on the size of the portal, as this has an influence on the distances to be covered

4.3.3   High-Bay Sorting Storage

In a high-bay sorting storage, the components are conveyed directly from the output of the cutting system into the working area of a storage and retrieval machine via roller conveyors and cross conveyors (Fig. 4.18). This is equipped with a lifting device, picks up the components and places them unsorted in free cantilever racks. The lifting device can also be equipped with individually controllable conveyor technology. It transports the components lying on top deeper into the compartments as needed and places several parts one behind the other. Components of different lengths are buffered and retrieved individually in this way. All other parts, which cannot be directly accessed due to chaotic buffering within a compartment, must be resorted. The storage and retrieval machine removes the materials in the required sequence. It then either provides the parts on conveyor systems for the subsequent process or for a handling device, which places the parts on load carriers. For plate materials, a vacuum sucker on the storage and retrieval machine is an alternative to a lifting mechanism. This is also capable of forming small stacks in the rack compartments.

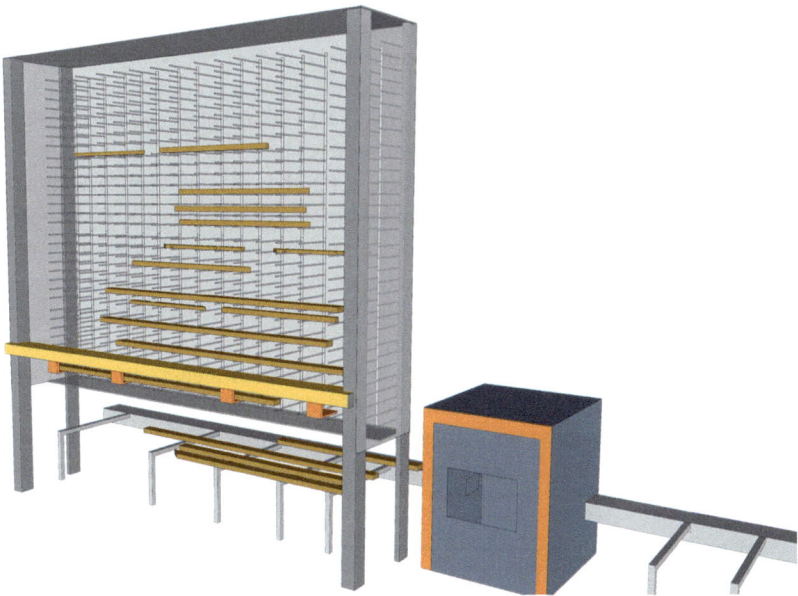

Fig. 4.18   High-bay sorting storage using the example of rods (analogous for plates)

Application (according to VDI 3415, sheet 1) per shift (8 h):	600-800 parts
Number of employees:	0 (only for monitoring, as automated process)
Note on application:	• If re-sorting processes become necessary, the performance of the system is reduced • the capacity (size) of the sorting bin has an impact on performance, as this affects the travel distances to be covered by the storage and retrieval machine

4.3.4 Robot Sorting Cell

Robots can take the components for sorting and buffering directly from the output table of the cutting system. If the adjoining placement is not possible, the parts are alternatively provided in the robot's working area via conveyor systems such as roller conveyors and cross conveyors (Fig. 4.19). The robot can be stationary or on a linear travel axis, thereby increasing the range and thus the capacity of the sorting cell. The pick-up can be done for materials with high density, for example, using vacuum suction cups, for porous materials using needle grippers, and for rods using jaw grippers. After the sorting process or buffering in the cell, the components are placed in the correct sequence for subsequent processes on conveyor technology or on load carriers. Since panels in timber construction often have notches, bevels, or other shapes, buffering the flat parts horizontally is more advantageous. Rods, on the other hand, can also be buffered vertically to save space. If several parts are chaotically combined in compartments, a resorting may be necessary when they are removed. The sorting process by a robot is unsuitable for long parts, as handling the large dimensions is only possible to a limited extent.

Application (according to VDI 3415, sheet 1) per shift (8 h):	600-800 parts
Number of employees:	0
Note on application:	• If re-sorting processes become necessary, the performance of the system is reduced • Shorter and smaller parts can be moved faster due to lower inertia and thus increase the possible output

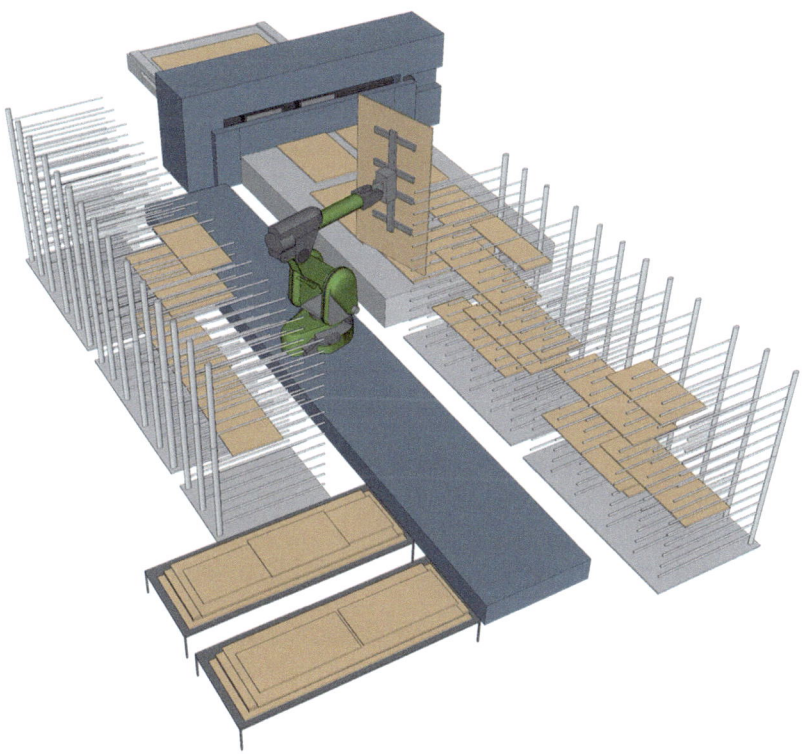

Fig. 4.19 Robot sorting cell using the example of panels (analogous for rods)

4.3.5   Comparison and Classification of Sorting and Buffering Systems

Digital systems for supporting manual processes, in conjunction with optimized load carriers, represent a sensible entry solution for small and medium-sized enterprises. For largely automatic processes, the described system concepts come into play. Particularly when considering the criteria of flexibility, capacity, and investment sum, clear advantages can be seen in a robot-based solution. Because it is also adaptable for future requirements (Table 4.6).

Table 4.6 Comparison of different sorting and buffering systems (evaluation of criteria from excellent (+++) to poor (− −))

Criteria	Digital Support	Area portal	High-bay sorting storage	Robot sorting cell
Capacity (batch size cutting)	−	+	+++	++
Flexibility connection to cutting	+++	++	+	++
Flexibility connection to subsequent process	+++	++	+	+++
Investment	+++	+	−	−
Future viability	+	−	+	++
Space requirement	++	−	++	+
Requirement for data	−	−	−	−
Process reliability	++	+	++	+

4.4   Internal Logistics and Material Provision

Many manufacturing companies focus on eliminating waste in value-adding processes and neglect the consideration of internal logistics and material provision. However, their optimal processes are essential for smooth production processes. The requirements for logistics and provision depend directly on the technologies used in component manufacturing. The decisive factor is whether the material supply between upstream and downstream processes is rigidly linked or decoupled. Since sheet materials are usually needed in several areas of production, a linked supply from cutting to component manufacturing is not feasibly realizable. A linkage of the rod cutting and the creation of bars, on the other hand, can prove to be efficient. In addition, the provision of components includes not only the previously processed materials but also other raw materials, semi-finished products, and purchased components such as windows. In principle, solutions should be sought that largely avoid the use of overhead cranes, forklifts, or pallet trucks. The time spent fetching these systems or waiting for their availability represents waste.

Fig. 4.20   Specialized Trolleys

4.4.1   Specialized Racks and Trolleys

Loading aids and provisioning methods optimally adapted to the different materials and components help to achieve waste-free processes in value-adding component manufacturing. According to the requirement of the transported goods as well as the later use and removal method, the loading aids are ideally developed individually according to Lean Management methods. The aim is always to increase efficiency by avoiding waste in the form of unnecessary handling and work steps (Ohno et al. 2013, p. 54). For a flexible design and use of transport means, it is advisable to avoid a large variance. Loading aids stored on rollers are moved manually (Fig. 4.20). Alternatively, the carriers are designed so that, for example, driverless transport vehicles can pick them up.

4.4.2   Individual Part Transport on Roller Conveyors, Belts, or Chain Conveyors

The individual transport of materials via various conveyor systems is particularly useful when processes are rigidly linked. However, this is only appropriate in areas with largely automated processes without repeated manual interventions. Because the connection of conveyor lines severely restricts access to the assembly tables (Fig. 4.21). A connection of processes through transport lines can be useful when certain parts are needed only at one or a few places. The provision should be practical without obstructing processes.

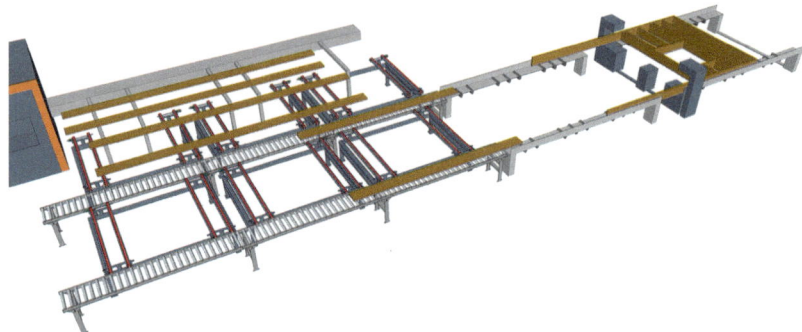

Fig. 4.21 Linking of Plants through Individual Part Transport using the Example of Rod
Cutting and Bar Frame Manufacturing

4.4.3 Driverless Transport Systems

Driverless transport systems (DTS) enable automated material movement through
driverless transport vehicles (DTV) (Fig. 4.22). The provision by DTV is either
controlled via a higher-level control system or initiated by employees of the
respective production areas. The potential of DTS can be exploited in production, particularly in connection with digitally managed buffer spaces. DTS ensure
that machines and workplaces are always supplied with material and that this
is reliably transported away and onwards after processing. For the transport of
different materials and components, there are numerous possible forms of DTV.
Their use ranges from lower load capacities such as the uptake of consumables
(e.g., screws) to the transport of stacks of cut materials to the movement of heavy
and bulky loads such as entire walls or loading bridges. For the efficient use of
DTV, they should therefore be adapted to the product to be transported. DTV
are offered with different steering structures and drives according to the desired
requirements. They are either line-mobile or area-mobile in their movement (VDI
2510 sheet 1 2009). There are also various types of navigation methods such as
following electrical guidelines or flexible laser navigation. In any case, it can be
ensured that collisions with the environment or employees are prevented.

Fig. 4.22   different DTV forms and transport possibilities through DTV

4.4.4   Comparison and Classification of Systems for In-house Logistics

For the area of transport and logistics, the combination of different systems is suitable in many cases. Automated solutions can severely limit the flexibility of some processes and usually require a lot of space. Therefore, they cannot be used sensibly in all areas. The choice also depends on the technologies used in the value-adding processes. For example, if robots are used in the assembly of the elements, the use of linked conveyor systems or AGVs should be considered. A human-free provision for robots simplifies safety technology and thus saves costs (Table 4.7).

4.5   Bar Assembly Manufacturing (Wall Elements)

In this manufacturing step, individual bar-shaped components are assembled into a framework. This is also referred to as the first step of pre-assembly of the two-dimensional elements.

Table 4.7 Comparison of different logistics systems (evaluation of criteria from excellent (+++) to poor (−−))

Criteria	Specialized Racks and Carts	Roller conveyors, Belt conveyors, Chain conveyors	Automated Guided Vehicles (AGVs)
Flexibility	+++	−	++
Investment	++	+	−
Future viability	+	−	++
Level of automation	−−	+++	+++
Space requirement	++	−	+
Data requirements	+++	+	−
Process reliability	++	++	+

4.5.1 Manual Tensioning Table for Wall Elements

A simple entry-level solution for the frame construction of wall elements is the manual tensioning table. It has cross and longitudinal tensioning devices that guarantee the construction of a right-angled element (Fig. 4.23).

Fig. 4.23 Manual tensioning table for frame construction, wall elements

Output (according to VDI 3415, sheet 1) or cycle output:	30-40 minutes per framing unit (approx. 10m length)
Number of employees:	2
Note on application:	• the output is heavily dependent on the number of employees in the process and the complexity of the boltwork • With a manual clamping table, it makes sense to use two employees

4.5.2   Frame Construction Station with Manual Insertion

The frame construction station with manual insertion is a common and widely used semi-automated solution in industrial standardized prefabrication (Fig. 4.24). While the bars are manually placed individually, the system mechanically advances the element in cycles (from post to post), so that the material is always inserted stationary. The posts are fixed in the machine by pressing stamps so that even twisted wood is positioned. They are then fastened at the ends with nails through the sill and rafter. In addition, the integration of other tools is optionally possible, such as a wave nailer, with which, for example, thick components can alternatively be fastened from above and below. In addition to the insertion of individual pieces of wood, it is also possible to introduce prepared window modules, thus reducing cycle times in the frame construction station. Due to the automated feed, the discharge area is fenced off for safety reasons.

Output (according to VDI 3415, sheet 1) or cycle output:	20-30 minutes per frame work (approx. 10m length)
Number of employees:	1
Note on application:	• the output is highly dependent on the complexity of the frame work and the preparatory activities outside the frame work station • If, for example, window modules are prepared outside the frame work station by another worker, the processing time in the station is reduced

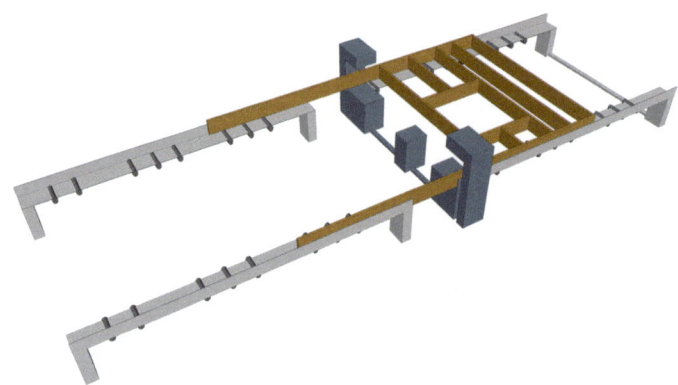

Fig. 4.24 Frame construction station with manual insertion

4.5.3 Frame Construction Station with Robotic Insertion

The degree of automation of the technology from Sect. 4.5.2 can be increased with the use of a robot, so that the processes are largely automated. Here too, the frame is automatically advanced, while the posts are inserted by an articulated arm robot with a gripping tool. The machine takes over the precise positioning of the parts and the subsequent fastening analogous to the system from Sect. 4.5.2. Manual interventions are still necessary in this design as soon as more complex structures such as window replacements need to be created. The protection of the employees is ensured, for example, by a movable fence (Fig. 4.25).

Output (according to VDI 3415 Sheet 1) or cycle output:	15-25 minutes per frame work (approx. 10m length)
Number of employees:	0,5
Note on application:	• the output is highly dependent on the complexity of the frame work and the preparatory activities outside the frame work station • for monitoring the system and for complex connections, a worker is required. However, this person can prepare window modules, for example, while the system is working automatically.

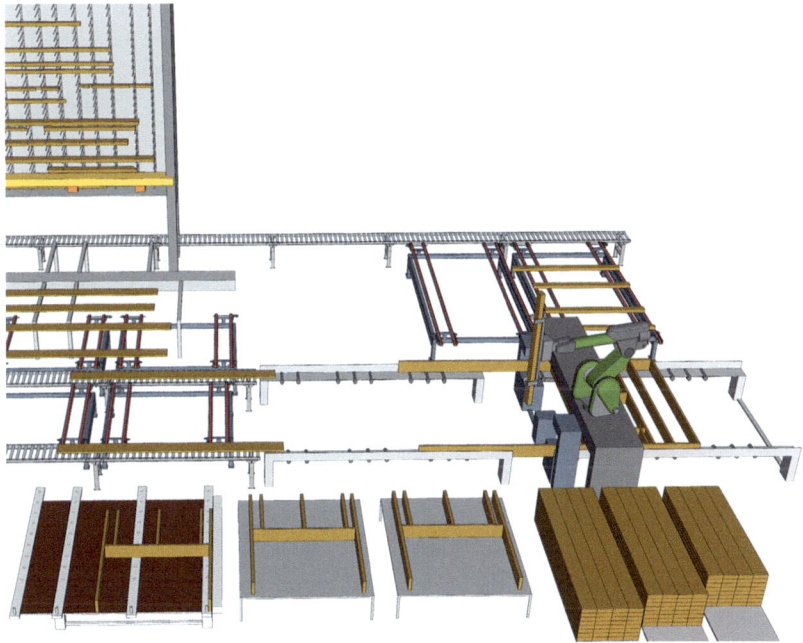

Fig. 4.25 Frame construction station with linked component transport from the cutting and insertion by robot

4.5.4 Portal-Articulated Arm Robot Combination

This stationary solution approach, in which all work steps are carried out without further transport of elements, is derived from the manual assembly table. In this case, the portal solution represents a fully automatic alternative complement to the articulated arm robot (Fig. 4.26). The rails on which portal robots move can either be embedded in the floor or lie on stationary supports at height (Fig. 4.26). The latter concept has the advantage that less dirt can accumulate there. Portal robots allow a greater range and higher load capacity compared to articulated arm robots. They can therefore be used more flexibly and also move larger components. Depending on the technology used and the manufacturer, portals can be combined with articulated arm robots.

Such a concept could be automated in stages. In the first step, it would be conceivable to roughly position the components using a portal. In the highest

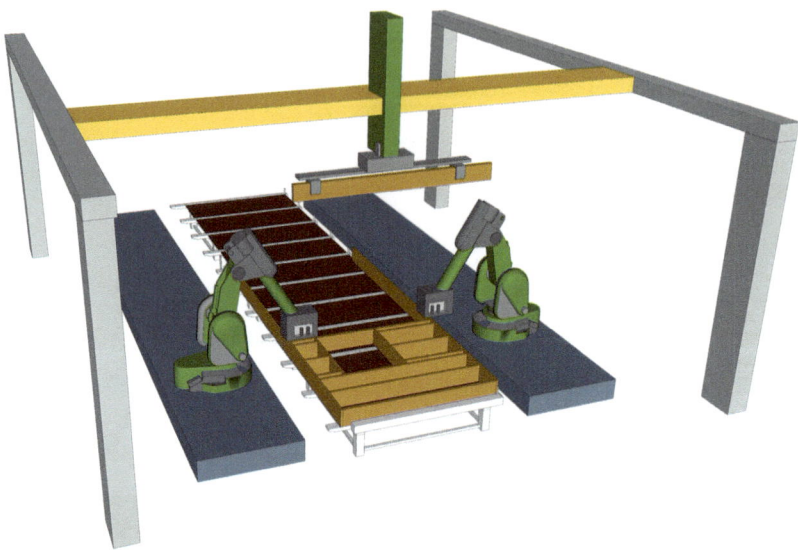

Fig. 4.26   Truss production by portal and articulated arm robots

expansion stage, the components could, for example, be precisely placed by a portal robot, fixed in the assembly table by a CNC-controlled clamping system, and fastened by articulated arm robots with appropriate tools.

Output (according to VDI 3415, sheet 1) or cycle output:	20-30 minutes per frame work (approx. 10m length)
Number of employees:	0 (only for monitoring, as automated process)
Note on application:	• the output is highly dependent on the complexity of the boltwork and the number of individual parts • This concept is designed to ensure that the process for all bolt work cards is fully automated and that employees only have to intervene manually in the event of faults

4.5.5 Comparison and Classification of Systems for Truss Production

The selection of the plant concept for the creation of a truss system strongly depends on the required performance and the demanded degree of automation. In a forward-looking company, a manual clamping table is only seen in the area of special production and not for the high-performance production of largely standardized elements. The described continuous truss workstations are a very efficient way to create truss systems. However, the portal-articulated robot combination offers the highest flexibility in terms of manufacturing possibilities among the solutions. Because in addition to simple truss systems, the production of more complex elements as well as the integration of further processing such as cladding is possible (Table 4.8).

Table 4.8 Comparison of different truss productions (evaluation of the criteria from excellent (+++) to poor (−−−)

Criteria	Manual Clamping Table	Truss Workstation manual	Truss Workstation Robot	Portal-Robot Combo
Performance (Processing Time)	−	++	+++	++
Flexibility	+++	+	+	++
Investment	+++	++	−	−
Future Viability	−	+	++	+++
Degree of Automation	−−	−	+	++
Space Requirement	+++	+	−	+
Data Requirement	+++	+	−	−
Process Reliability	++	+	−	−

4.6   Framework for Roof and Ceiling Elements

Due to the different tension direction and larger dimensions and lengths of the individual components, the frame production for roof and ceiling is considered separately. Essential here is the holding and tensioning of the statically load-bearing components such as rafters and ceiling beams parallel to the x-axis of an element table. The angular holding is usually ensured by a stop in the y-direction. The production of roof and ceiling elements takes place stationary without further transport, as the component tensioning and holding in the process can only be ensured with great effort. The connection and fixation of the components often only takes place with the cladding by panels or slats.

4.6.1   Manual Tensioning Table for Roof/Ceiling Elements

Analogous to the system for wall production (Sect. 4.5.1), there is also the manual tensioning table for the frame production of roof and ceiling elements (Fig. 4.27). The transverse and longitudinal integrated tensioning elements and stops are manually positioned and fix the timbers.

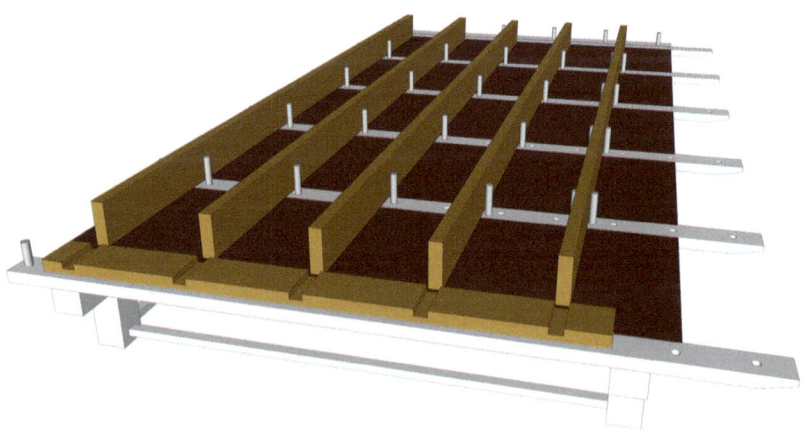

Fig. 4.27 Manual Tensioning Table Roof/Ceiling

Output (according to VDI 3415, sheet 1) or cycle output:	20-40 minutes per framework (approx. 2.5 m wide)
Number of employees:	2
Note on application:	• the output is heavily dependent on the number of employees in the process, the complexity of the element (replacements, positioning planks, etc.) and the effort involved in retooling the clamping elements • With a manual clamping table, it makes sense to deploy 2 employees

4.6.2 CNC Tensioning Table with Manual Insertion

CNC tensioning tables (Fig. 4.28) are similarly constructed as manual tables (Sect. 4.6.1), however, the stops and tensioning elements are set by the numerically controlled machine. This happens automatically based on control commands from CAD/CAM data, which are translated into motion sequences, eliminating manual setup and measurement. The long rafters and ceiling beams are manually positioned on the assembly table using a hall crane or separate lifting device and are held there by the pneumatically movable tensioning elements. The stop along the y-axis ensures the angularity of the elements. The integration of the tensioning tables into a production line is optionally possible by connecting them via transverse or longitudinal transport.

Output (according to VDI 3415, sheet 1) or cycle output:	15-30 minutes per framework (approx. 2.5 m wide)
Number of employees:	2
Note on application:	• the output is highly dependent on the number of employees in the process and the complexity of the element (replacements, screeds, etc.). • Two workers are required to insert and align the large components • With this variant, set-up times are greatly reduced so that employees can concentrate on value-adding processes

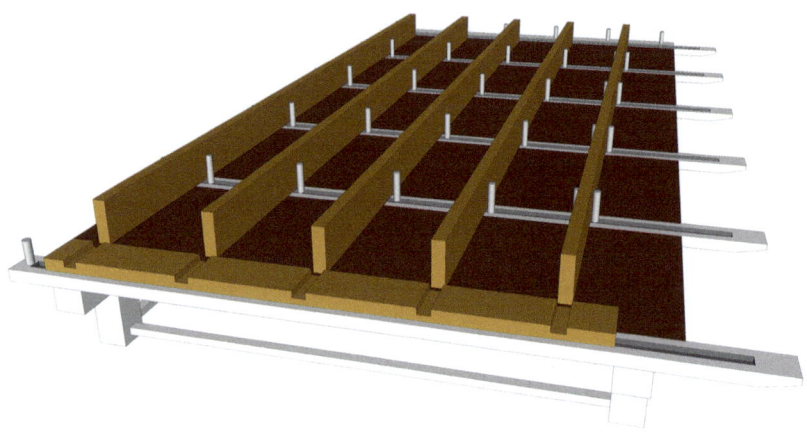

Fig. 4.28 CNC Tensioning Table Roof/Ceiling

4.6.3 CNC Tensioning Table with Automated Insertion

The manufacturing method described in Sect. 4.6.2 using a CNC tensioning table can be expanded to automate the insertion and, if necessary, fastening of the timbers. Depending on the equipment, the processes can be partially automated or fully automated in the form of a cell (see Sect. 4.5.4) (Fig. 4.29).

Output (according to VDI 3415, sheet 1) or cycle output:	15-30 minutes per framework (approx. 2.5 m wide)
Number of employees:	0-1
Note on application:	• The aim of this solution is not to increase cycle times, but to reduce employee deployment.

4.6.4 Comparison and Classification of Systems for Frame Manufacturing Roof Ceiling

While manual tension tables represent a good entry-level solution with low investment, the setup of the tensioning elements, especially with high variance, consumes a lot of time. A CNC tension table, with its manageable complexity,

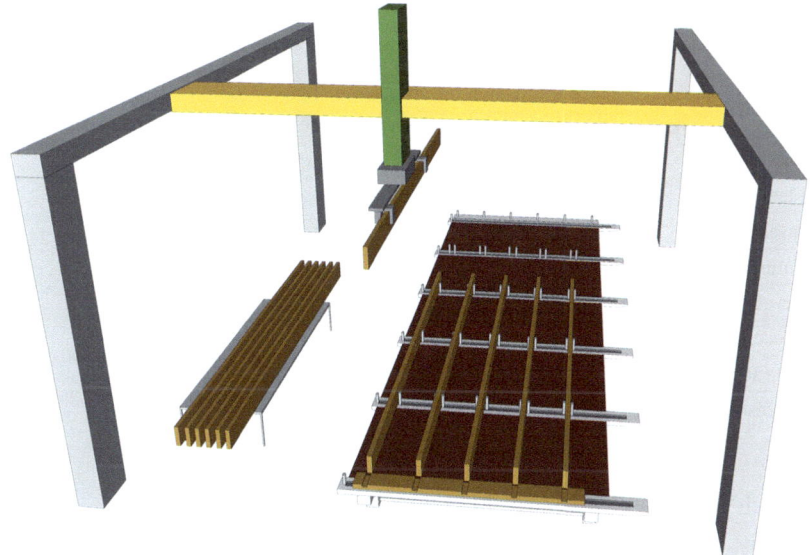

Fig. 4.29 CNC Tensioning Table Roof/Ceiling with Portal

provides a practical solution for small and medium-sized companies. Automated insertion is considered when the personnel deployment for scalable performance improvement is to be reduced (Table 4.9).

4.7   Laying of the Cladding

A crucial part of prefabrication in timber construction is the cladding of load-bearing post or frame structures, cross-laminated timber or solid wood elements with panel materials. Since the individual cladding layers fulfill different functions, various mineral and wood-based materials are used, which are designed for the respective requirements (Table 4.10). The type of cladding execution, fasteners, and seals of the levels must also meet the respective conditions. If one also considers the requirements of a circular economy, in which the components are reused or recycled at the end of their life, the fasteners must be chosen appropriately.

Table 4.9 Comparison of different frame manufacturing systems Roof/Ceiling (Evaluation of criteria from excellent (+++) to poor (− −))

Criteria	Manual Tension Table	CNC-Tension Table	CNC Tension Table with auto. Insertion
Performance (Processing Time)	−	++	++
Flexibility	++	+	+
Investment	+++	+	−
Future Viability	−	+	++
Level of Automation	− −	+	++
Space Requirement	++	++	+
Data Requirement	+++	+	−
Process Reliability	++	+	−

Table 4.10 Different functions of the cladding layers and possible materials

Functions of the cladding layers	Example materials (in plate form)
Room-side finish	Gypsum board or gypsum fiber board
Installation level	Softwood fiber with milled grooves and cutouts for installations
Bracing	Gypsum fiber board or OSB3
Vapor barrier	OSB3 or wood fiber boards
Second water-bearing levels	Softwood fiber board
Façade	Wooden formwork or cement fiber board

The panels are pre-cut and processed individually for the respective element during the cladding process, depending on the production concept, or are applied in their entirety. For the latter, large-format panels are often used, which span the entire element height and several compartments (panel width ≥125 cm). The advantage is that fewer individual parts need to be moved and fastened, saving process steps and time. However, since these panel formats are difficult to handle through automation, they are mainly used in the craft environment.

4.7.1 Manual using Vacuum Suction Cups or Needle Grippers

With vacuum suction cups, panels can be sucked in so that they can be easily moved manually. For air-permeable materials like some softwood fiber boards, there are alternative systems like needle grippers that engage with the material. The handling device is flexibly attached to either a swing arm crane or an overhead crane system. While the swing arm is a simple and cost-effective solution, the overhead crane is easier for employees to operate. Due to the linear movements, the panels are easier to position (Fig. 4.30). The aim of these systems is to improve ergonomics and facilitate manual laying. The risk of panels being damaged during handling is minimized.

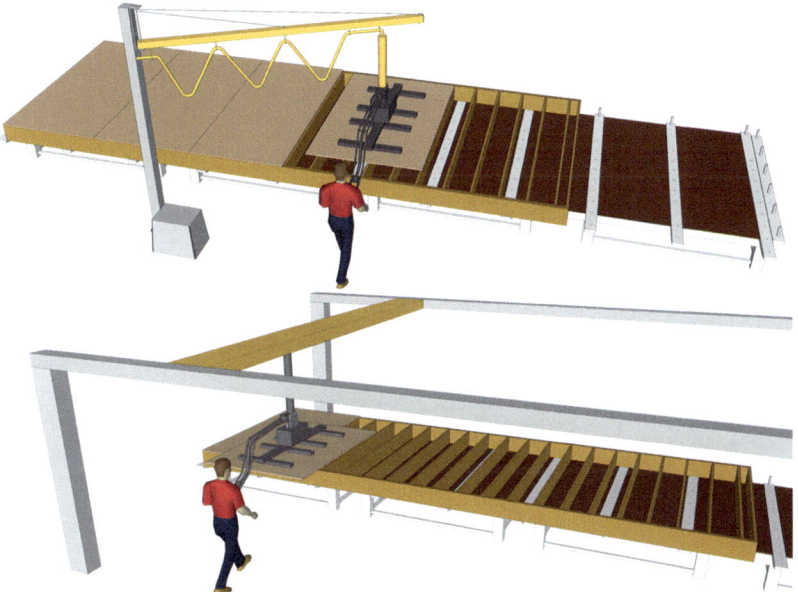

Fig. 4.30 Laying of cladding by manual handling devices

Output (according to VDI 3415, sheet 1) or cycle output:	10-15 minutes per panel layer (approx. 10m panel length)
Number of employees:	1
Note on application:	• The output depends heavily on how many individual panels have to be placed and to what extent the worker has to sort pre-cut panels.

4.7.2   Remote-Controlled Manipulators

Through remote-controlled rigid manipulators, employees can move and position panels without manual intervention (Fig. 4.31). The materials are picked up by attached vacuum suckers or nail grippers, which are moved by motor drives such as scissor or telescopic hoists. A rigid hoist facilitates panel handling, as the panel, even if it is not picked up exactly at the center of gravity, does not tilt significantly and can be moved horizontally.

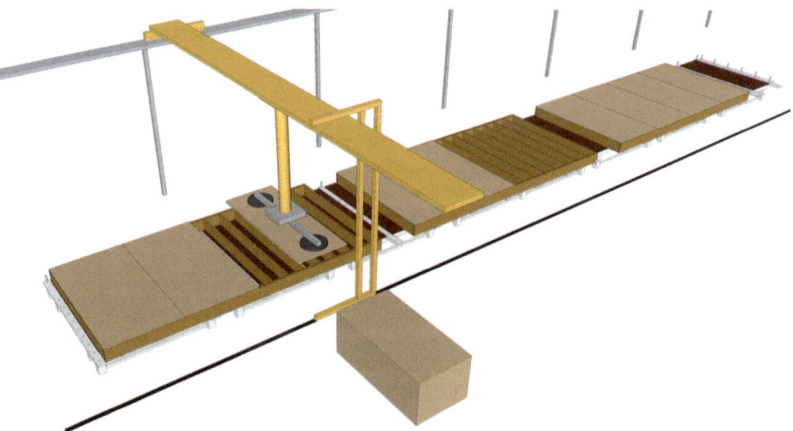

Fig. 4.31   Panel placement by remote-controlled manipulators

Output (according to VDI 3415, sheet 1) or cycle output:	10-15 minutes per panel layer (approx. 10m panel length)
Number of employees:	1
Note on application:	• Although this method is slower when placing the individual panels, large-format panels can be moved with a high degree of work safety, as the operator can keep a sufficient distance.

4.7.3   Articulated Robot with Handling Unit

By using an articulated robot, the panel placement process can be fully automated (Fig. 4.32). The panels are picked up and moved by the robot using the aforementioned handling tools. A prerequisite for the correct pick-up and subsequent positioning of the components is the detection of the zero points and orientation. This referencing can be done by placing the panel on a tilted table, where the panel slides into a corner stop. However, this requires the panel to be picked up and put down twice. Alternatively, camera systems can be used that scan and visually recognize the component. Once the zero point and orientation have been identified, the part can be precisely picked up and maneuvered to the predetermined position on the element. The step of referencing is essential for the alignment of the components. Because a challenge in using robots is maintaining tolerances. Without accurate component recognition, for example, too large gaps can occur or panels can be applied overlapping.

The components can be placed sequentially from one side of the element to the other or in any order. When working in sequence, the previous panel can be approached and used as a stop, which is useful for more accurate positioning. The first panel is stapled after placement to prevent slipping. The following parts are initially only placed and then fixed.

Fig. 4.32 Panel placement by articulated robot

Output (according to VDI 3415, sheet 1) or cycle output:	10-15 minutes per panel layer (approx. 10m panel length)
Number of employees:	0 (only for monitoring, as automated process)
Note on application:	• the output depends heavily on how many individual panels have to be laid down • Non-productive times also have a strong influence on performance; they are caused, for example, by referencing, the slow removal of panels from a stack to prevent further panels from being removed or slipping, or the travel path of the robots on the linear axes

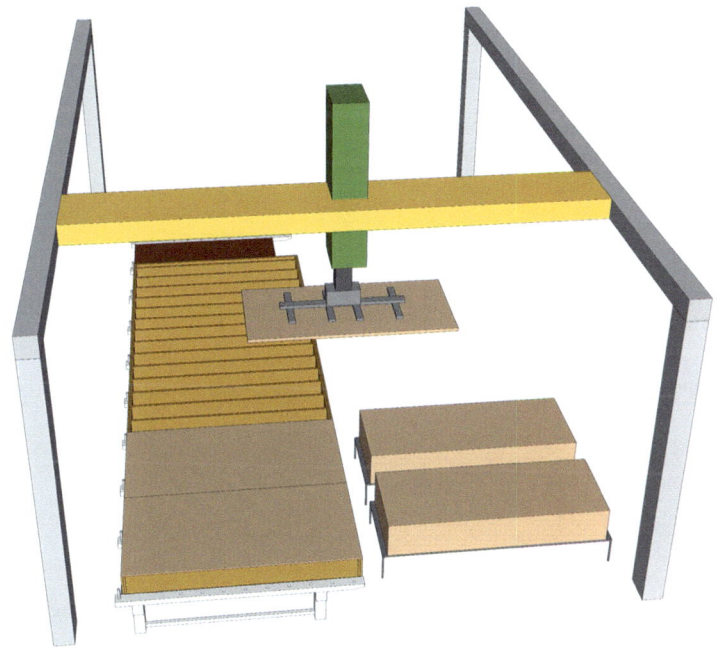

Fig. 4.33 Portal robot with elevated rails

4.7.4 Linear or Portal Robots with Handling Units

A fully automatic alternative to the articulated robot is a portal solution with a larger range and higher load capacity (Fig. 4.33) compared to articulated robots (see Sect. 4.5.4).

Output: similar to Sect. 4.7.3.

4.7.5 Comparison and Classification of Systems for Panel Placement

The laying of the plates is still often done manually, but can be relatively reliably automated. The greatest challenge is the referencing of the plates, which is a prerequisite for maintaining tolerances in positioning. A remote-controlled

Table 4.11 Comparison of different systems for laying the planking (evaluation of criteria from excellent (+++) to poor (−−))

Criteria	Manual using vacuum suction or needle gripper	Remote-controlled manipulators	Articulated arm robot	Linear or portal robot
Performance (processing time)	++	++	++	++
Flexibility	++	+	+	++
Investment	+++	++	−	−
Future viability	−	+	++	++
Degree of automation	−−	+	+++	+++
Space requirement	++	++	+	+
Data requirement	+++	+	−	−
Process reliability	++	+	+	+

manipulator is a sensible partial automation for the supported handling of large-format plates. For the production of average element dimensions, the articulated arm robot is a fully automated solution with sufficient flexibility. However, since the reach of an articulated arm robot is limited, a portal system is particularly useful when larger element dimensions (element width ≥3.5 m) are produced (Table 4.11).

4.8   Attaching and Processing Cladding

Once the plate-shaped materials have been placed on the element, the parts are attached and processed. Depending on the requirements of the material and the function of the respective cladding layer, different fasteners are used. Common types of fastening include, for example, clips, nailing, screwing, or gluing. The most important and most frequently used fasteners are clips. They can be inserted manually or automatically in appropriate dimensions for different cladding materials in the shortest possible time. The distance between the individual clips, the clip spacing, varies depending on the function of the cladding layer.

After fastening, if this has not been done beforehand, the plates are cut and processed for the final formatting, cutouts, drilling, and milling.

4.8.1   Semi-Automated Clip Device

A simple entry-level solution for automating clipping is a semi-automated pneumatic device (Fig. 4.34). It features, for example, positioning aids and guiding devices such as rails, which ensure optimal clip alignment and adherence to the specified edge distance. The devices can be moved both manually and via a self-propelled cart or rail systems. The adjustable clip spacing allows the clips to be inserted in the defined area. While semi-automated solutions for clipping can be useful, manual cutting on the element should be avoided as waste handling and extraction are too complex for industrial production. Semi-automated clip devices are often used for the pre-assembly of special elements and complex components.

Fig. 4.34   Attachment using a semi-automated clip device

Output (according to VDI 3415, sheet 1) or cycle output:	15-20 minutes per panel layer (approx. 10m panel length)
Number of employees:	1
Note on application:	• As wall, roof and ceiling elements are constructed differently, the number of fixing points and therefore the number of staples per square meter varies depending on the element type. This has an impact on the application of the staplers.

4.8.2   CNC Processing Portal with Tools

A widely used solution is the CNC-controlled processing portal, which is guided by rails in the floor and equipped with different tools (Fig. 4.35). The floor rails allow a travel path across several workstations. Tool changing systems integrate several units into one portal. In addition to fastening devices, tools for machining can also be used, such as milling spindles, sawing units, and drilling heads. Depending on the design of the system, it can also take over the processing of solid wood elements. However, these processes add the challenge of extracting dust and chips. This must be moved with the portal and be able to separate wood and mineral materials. To minimize the walking distances of employees during manual tasks, the width of the portal (travel path that must be free next to the processing tables) should be kept as small as possible.

Output (according to VDI 3415, sheet 1) or cycle output:	8-15 minutes per panel layer (approx. 10m panel length)
Number of employees:	0 (only for monitoring, as automated process)
Note on application:	• the application refers to the fastening process using a stapling unit on an element • it depends on the element length, the bracket length and the bracket spacing • If screws are used for fastening, it also depends on whether several magazine screwdrivers are connected in parallel or whether only one screw is inserted for each fastening process

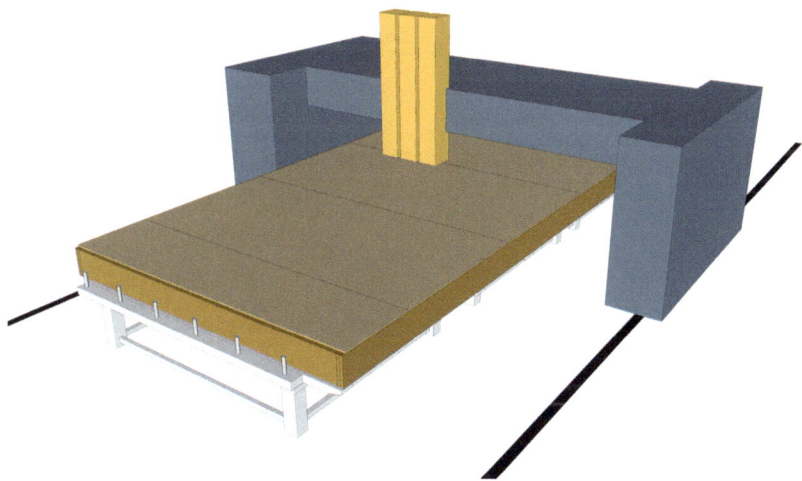

Fig. 4.35   Attachment of cladding by a CNC portal

4.8.3   Articulated Robots with Tools

The articulated robot is also a flexible automation solution for the attachment and processing of cladding (Fig. 4.36). Since units such as plate suckers, nailers or screwdrivers can be easily changed on the robot, it is usually used for both the laying as well as the attachment and processing of the plates. If machining operations are carried out, however, the extraction of dust and chips must be ensured, as well as the safety-related housing of the tools. To increase the reach of the robot, it is mounted on linear axes, which can be located on one or both sides of the processing table.

Output: similar to Sect. 4.8.2, as the limiting factor is the trigger speed of the stapler and not the feed speed of a robot or portal.

4.8.4   Linear or Portal Robots with Tools

Linear and portal robots can be used similarly to the articulated robots (Sect. 4.8.3), however, their machine frame allows them a greater range and load capacity for larger and heavier tools.

Output: similar to Sect. 4.8.2.

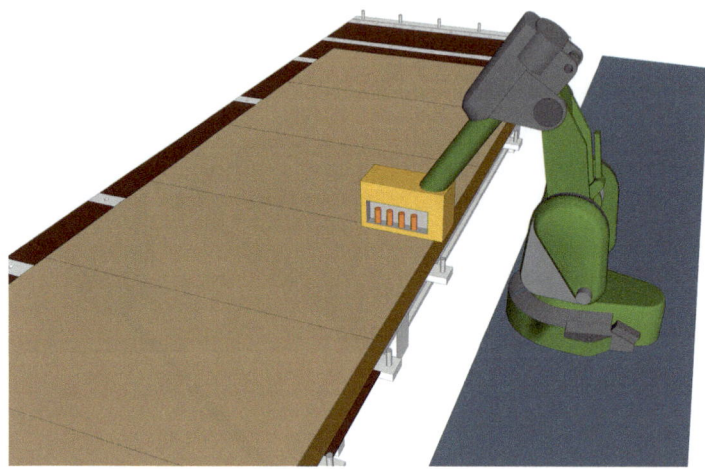

Fig. 4.36 Attachment of cladding by articulated robot

4.8.5   Comparison and Classification of Cladding Attachment and Processing Systems

When manually attaching the plates, particularly maintaining the specified clip spacing to the edge of the plate is challenging, which leads to increased processing times for precise positioning. The process times of the automated solutions are all similarly estimated, as all technologies use very similar aggregates and the processing is generally data-based (Table 4.12). The decision for a system is usually made in connection with the technology selection for plate laying, so that the degree of automation of both processes matches or the systems perform both processes.

4.9   Turning

Since the elements are usually created lying on worktables, it is necessary to turn them for double-sided assembly and processing. The process typically follows the completion of the first side of the cladding (usually on the room side) in order to insulate the elements and clad them again. Although turning is not a value-adding process, it still influences the processing time of the elements. Especially when

Table 4.12 Comparison of different fastening systems for cladding (evaluation of criteria from excellent (+++) to poor (−−))

Criteria	Semi-automated clip device	CNC processing portal	Articulated arm robot	Linear or portal robot
Performance (processing time)	−	++	++	++
Flexibility	++	+	++	+++
Investment	+++	++	+	−
Future viability	−	+	++	++
Degree of automation	−	++	++	++
Space requirement	++	++	+	+
Data requirements	+++	−	−	−
Process reliability	++	++	+	+

adhering to cycle times of a line production, the process can become an obstacle. In order to at least be able to use the turning area productively, activities such as the insertion of pre-installations of the electrical system or insulation work are often added there. For safety reasons, the turning is usually supervised and carried out by employees. Full automation of the process is only possible under increased safety requirements (e.g., fenced cell).

4.9.1 Turning using Overhead Crane or Manipulator

If there is no defined turning station, turning using an overhead crane or manipulator is the simplest option (Fig. 4.37). The elements are, for example, attached at hanging points, which are anyway needed for the later movement and assembly of the elements by a crane on the construction site. The use of a crossbeam optimizes the load distribution. When the element is lifted, it should be noted that stresses present in the element can possibly lead to damage. Impact-sensitive materials such as plasterboard can also quickly be affected. Overhead cranes and

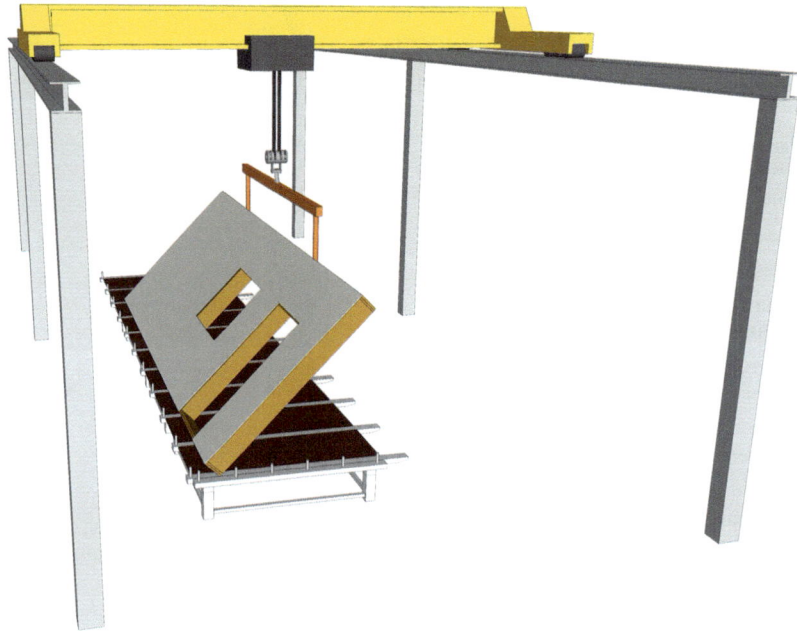

Fig. 4.37 Turning using overhead crane

manipulators are predominantly used in the craft sector for turning and are usually only found in industrial productions at special production sites.

Process duration:	2-5 minutes
Number of employees:	1
Note on the duration of the process:	• The duration depends on the extent to which preparatory activities such as fetching a crane, waiting for availability or preparing the suspension points are involved.

4.9.2 Turning through Two Element Tables

A common solution is turning on two parallel element tables (Fig. 4.38). The element lies on the so-called giver table, from where it is transferred to the receiver table. For this, one table moves onto the other or both tables move towards each other at the same time, after which they are hydraulically erected. Subsequently, the element, which is now standing upright on brackets attached to the table, is tipped from one table to the other. There are lifting devices for detaching the elements from the giver table. Further equipment such as clamping devices and transport devices can be integrated into the turning tables. Thus, their use is also possible for element production and further transport in a line. Furthermore, two turning tables can form a self-contained cycle production, in which one side of the element is produced in the first cycle on the giver table and the second side of the element is completed after turning in the second cycle on the receiver table. While the completion is running on the receiver table, the production of a new element can already begin on the giver table. Before the next element can be turned, the receiver table must be empty. The use of two element tables for turning results in the production line requiring two parallel rows of tables in any case (staggered production line).

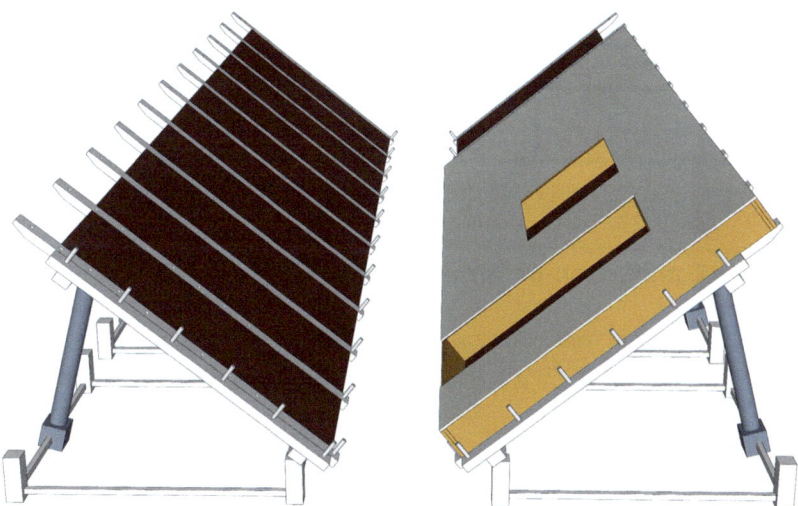

Fig. 4.38 Turning through two element tables

Process duration:	approx. 2 minutes
Number of employees:	1
Note on the duration of the process:	• The duration depends on the extent to which preparatory activities, such as conveying the element or moving the donor or recipient table, are still necessary.

4.9.3 Turning on the Spot

An approach that has only established itself in recent years is turning on a single table or work area (Fig. 4.39). This solution is mainly used in conjunction with a long, continuous element table due to manufacturer specifications (see Sect. 3.3.2), but it can also be used as a single component of a cycle line with several individual processing tables. Depending on the length of the elements to be manufactured and turned, several turning devices are used. This type of turning allows the layout design to arrange manufacturing tables in a single line without the need for an offset as when turning through two element tables. This circumstance reduces the space requirement of the manufacturing tables.

Process duration:	approx. 2 minutes
Number of employees:	1
Note on the duration of the process:	• The duration depends on the extent to which preparatory activities, such as conveying the element, are still necessary

4.9.4 Comparison of Turning Systems

Turning on the spot without the use of two tables allows for a leaner layout with the straight production line. It also offers the advantage that the process is not dependent on the status of a downstream table or area. Before turning through two element tables, the receiving table must have been emptied, which is why no value-adding activities are possible during the turning time on both tables. On the other hand, if turning is done on the spot, the work of the adjacent areas can continue in the meantime. The subsequent transport of the elements takes place

Fig. 4.39 Turning on the Spot

simultaneously for a more homogeneous production flow. In addition, this turning technology allows the workplace to be designed as a closed island production.

4.10 Insulation of the Timber Frame Compartment

In the area of insulation, there has been a strong further development of the offer and a clear trend towards renewable insulating materials such as softwood fibre, cellulose, hemp, straw etc. in recent years. Various materials in different forms are still used. Flat insulation is used, for example, in thermal insulation composite systems analogous to panel processing. Many materials are also compressed into mats or panels. Since they are often not dimensionally stable, their insertion into the compartments cannot usually be automated with process reliability.

In contrast, there is loose insulation material that is blown into the compartments of timber frame construction elements. The insulation process of blowing in is on the rise, as it is very well automatable. In addition, there is the significant advantage that no material losses occur due to cutting waste. This eliminates the costs and the process of disposal as well as the associated space requirement and logistics effort. This so-called blown-in insulation can be introduced both before and after the cladding of the elements. This is referred to as insulating into open or closed compartments.

Blowing into closed compartments has been in use on construction sites for some time. The prefabricated elements are clad on both sides and not insulated. Openings are made in the room-side cladding layer (panels or films) in prefabrication or on the construction site, through which the insulation is then blown in via hoses after assembly on the construction site. This process can also be carried out analogously in prefabrication. However, since the distribution of the insulating material in the closed compartment cannot be visually monitored, the insulation in prefabrication is largely blown into open compartments for quality control reasons. For this purpose, there are blow-in plates that are positioned over the compartment to be insulated and blow in the material over a large area through several nozzles. Nevertheless, controlling the volume of insulation material also poses a challenge when blowing into the open compartment. The volume can be determined either by the volume flow and the blow-in time. Alternatively, a weight check can be carried out using a weighing function in the processing table by weighing the element after each compartment has been blown out and thus determining the difference.

The insulation material is usually delivered in bags for small quantities, and in large bales for large quantities, which are loosened into blow-in insulation in bale mills. While the manufacturers of the insulation materials usually also offer the systems for blowing in, they only provide the necessary automation to a limited extent. Independent machine builders usually have to be consulted for this.

The following will shed light on the automation possibilities for blowing insulation into open compartments in prefabrication.

4.10.1 Manually Guided Blowing Plate on Crane or Portal

A semi-automated entry solution that can be used in both craft and industrial production environments is the blowing plate attached to the portal, hall, lightweight or wall-running crane (Fig. 4.40). It is manually guided by one to two employees and positioned on the compartments to be insulated. The control is done directly on the plate.

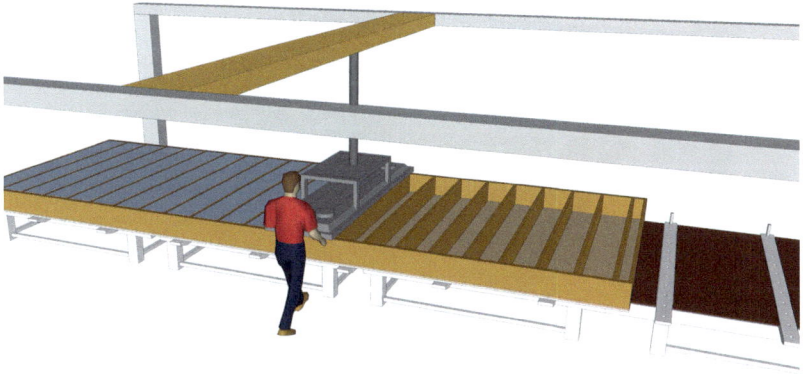

Fig. 4.40   Insulating with manually guided blowing plate

4.10.2   CNC Machining Portal with Insulation Plate

The already mentioned system of the CNC machining portal can also be used for blowing in the insulation by attaching an insulation plate (Fig. 4.41). Since it receives the information about the location and volume of the areas to be insulated via CNC data, the insulation plate positioning and blowing process are carried out without the intervention of employees. With the possibility to rotate the insulation plate by 90°, even longitudinally lying compartments can be insulated by the portal. If the CNC machining portal is not fully utilized by the insulation process, it can possibly also be used for upstream or downstream processes with additional tools.

4.10.3   Linear or Portal Robot with Insulation Plate

Linear or portal robots, which are used for the creation of the framework or the cladding layers, are also capable of taking over the insulation process fully automated (Fig. 4.42). The systems can be designed specifically for the insulation process or so that the robot can switch between the insulation plate and other units such as plate suckers and thus perform several processes. This allows for cell production, where all processes are carried out at one location.

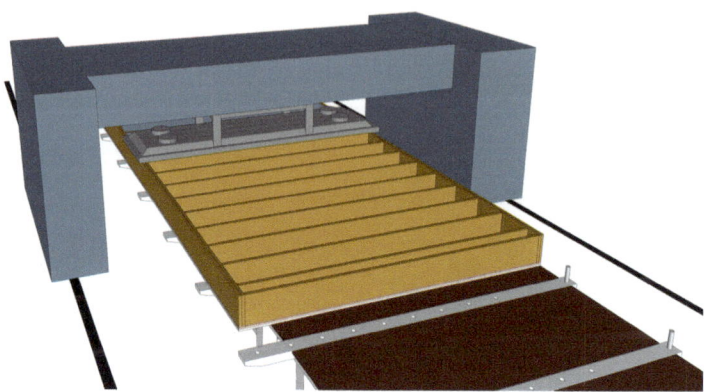

Fig. 4.41 Insulating with blowing plate on CNC portal

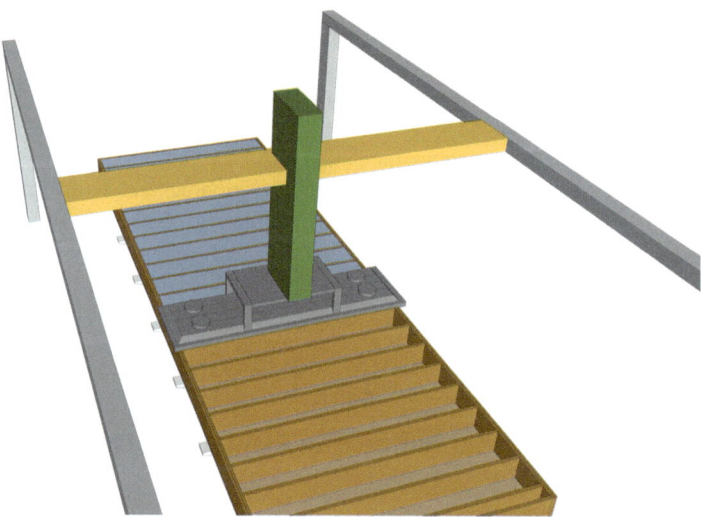

Fig. 4.42 Insulating with blowing plate on portal robot

4.10.4 Comparison of Systems for the Integration of an Insulation Plate

The systems described above usually differ only in flexibility, degree of automation and the required investment height. Considering these factors, the CNC machining portal is in most cases to be considered as a sufficiently flexible and cost-effective solution. For industrial productions, the manually guided blowing plate requires too high a personnel deployment.

4.11 Plastering

Although the trend towards renewable raw materials can also be found in the design of building facades and therefore increasingly relies on wooden formwork, plastering remains relevant, particularly in single-family home construction. Some productions integrate the plastering of the exterior wall elements in their finish area of prefabrication after the completion of the double-sided element cladding. For the application of plaster, the outer cladding consists of a plaster base plate, which is available in different materials such as EPS, rock wool or softwood fibre boards.

In prefabrication, usually only the reinforcement plaster, which consists of reinforcement mesh and undercoat, is applied. To avoid damage to the surface that can occur during transport and assembly, the top coat of plaster is only applied on site. This also makes it possible to seal the joints between the elements for a visually more appealing result after assembly.

The automation of the process is possible, for example, by using a portal that moves over the stationary element, unrolls the reinforcement mesh and applies and levels the plaster. For an alternative solution, the element is moved standing through a stationary plaster application machine, which unrolls the reinforcement mesh and sprays the base plaster in vertical movements. While the application of interior plaster can already be carried out by robots, the use for exterior plaster has so far been avoided, which is particularly related to the different plaster material and its higher degree of contamination.

Although there are various approaches to the automation of plastering, there are no mature standard solutions on the market. Systems of this kind are usually designed individually for productions in cooperation with mechanical engineers and plaster manufacturers. Despite automated plaster application, the overall process is still associated with many manual tasks, such as at window reveals.

Thus, there is a tremendous need for action for the research and development of automated plastering solutions.

4.12 Battening

Battens are used in timber construction, for example, as load-bearing and sub-structure for roof covering and facade elements or also on the room side in the ceiling and roof area. Depending on their respective function, they must be aligned longitudinally or transversely on the element. The decisive advantage of automation in this area is that the often complex measuring activity, component positioning, alignment and fastening runs automatically. With appropriate tools, articulated arm and portal robots can take over the application of the battening. For the alignment of the battens transversely to the element, the process can also be efficiently automated by using a CNC machining portal in connection with a travelling batten magazine, which can be loaded outside the plant. However, if the battens are aligned longitudinally on the element, the handling of the long components by the machining portal is usually not possible due to dimensions. Usually, the automation solution is chosen with a focus on the upstream and downstream processes and equipped with the necessary tool, so that the laying of the battening can also be taken over.

4.13 Formwork

For the application of formwork boards, there are initial automation approaches through robots, which, however, have not yet been widely used. In manual attachment, the boards are usually only finally formatted after fastening to facilitate the process of positioning and alignment. In this way, the highest precision is not required in these processes, as the final dimension is either manually or with the help of a CNC machining portal finally produced on the element. When using a robot, the cutting of the boards can be upstream, as the automatic positioning is very precise without extending the process times. Window reveals, however, are carried out manually regardless of the selected automation solution or solved by prefabricated window modules. Depending on the execution of the formwork, a CNC machining portal can also be used for fastening in addition to the process of laying, if the fasteners do not have to be hidden.

4.14 Further Digital Support Options

The more complex the construction of the elements, the more difficult it is to automate the manufacturing process. To make even manual processes as efficient as possible, Lean Management tools are used, which are based on the Toyota production principle. The basic idea is to eliminate waste in all processes (Ohno et al. 2013, p. 35). This is achieved on the one hand by considering individual tasks and skills of the employees and on the other hand by teamwork to meet the goal of deadline compliance (Ohno et al. 2013, p. 41). Seven types of waste are distinguished: overproduction, waiting times, transport, ineffective processing, storage, unnecessary movements, and defective products (Ohno et al. 2013, p. 54). These must be identified and reduced and eliminated with the right tools.

One solution is the digitization of processes. A first step for the production area is the digital provision of plans for the production staff. Because paper construction plans do not contain any metadata that reveal detailed information and do not allow interaction to call up further details about a component if necessary. To update the plans after adjustments have been made, they must be reprinted and manually distributed. The use of digital assistance systems makes it easier for employees to access data and construction plans. This requires extensive data collection and processing in the areas of construction details and implementations of all processes. Waste and downtime caused by unclear or outdated plans are thus reduced.

4.14.1 Touch Displays and Industrial Tablets

One tool for the digital provision of information in manufacturing is touch-sensitive monitors or tablets. The plans are automatically updated in this way and can be used interactively, so that employees can display specific details as needed.

4.14.2 Laser Projection

In particular, to make the time-consuming tasks of reading plans and measuring component positions obsolete, it is possible to use laser projectors that project the component positions from the construction data onto the assembly table (Fig. 4.43). The projectors are capable of displaying all layers of the element, each individual component, and even all attachment points and processing areas

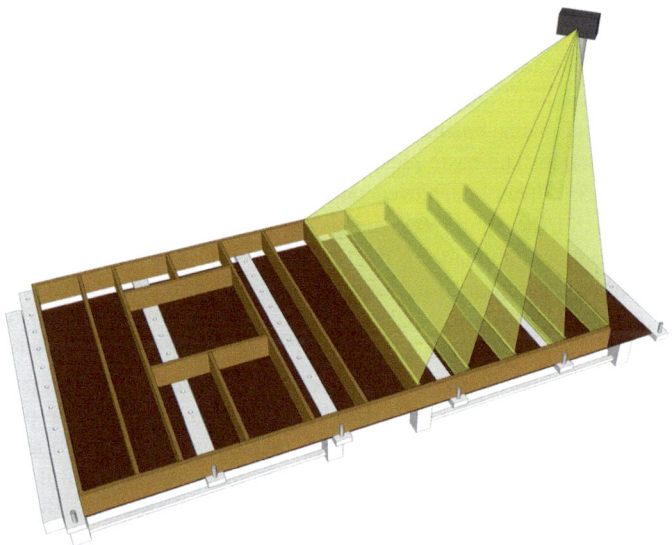

Fig. 4.43 Laser projection for post and beam construction

for cuts or milling. Depending on the design of the device, the individual cat-
egories can be displayed in different colors. The laser systems can display the
component outlines with tolerances below 1 mm. It is possible to display the
projection layer by layer or the individual components step by step in assembly
order. The display change by the projector takes place either via manual feedback from the employees or through camera systems that recognize the correct
positioning of the components and report this to the laser system. The use of
these systems and the way the laser display is used depends on the detail and
data preparation.

4.14.3 Mixed Reality

Mixed Reality uses a similar approach to laser projection by visualizing information for employees. The main component is a pair of glasses that displays the
elements with the individual components in a virtual reality three-dimensionally
in the middle of the real production environment. This allows employees to see
the construction to be created virtually on the assembly table and to build it

up successively. This technology thus offers extended options for information provision compared to laser projection.

4.14.4 Pick-by-Light and Pick-to-Light Systems

The Pick-by-Light and Pick-to-Light systems support employees, for example, in picking small parts in the warehouse, components after cutting, and in assembling the elements, by giving them light signals. This makes it clear to them which components are to be placed in which areas or which parts are to be picked up. This saves them search times and unnecessary handling due to the resorting of parts. A prerequisite for the usability of this technology is suitable data creation including sorting or production sequence.

Digital Process and Information Flow in Timber Construction

5

In considering the automation solutions described above, it is unmistakable how significant a role extensive data generation, processing, and provision plays. For the implementation of fully automatic processes, it is essential to create data from design plans and details for all materials and components that contain the necessary information for all processes.

However, a waste-free production process is only achievable if this data is complete and error-free. This applies not only to prefabrication in timber construction, but rather to the entire construction process. Since many different responsible departments and trades are part of the overall process chain from the beginning of planning to construction acceptance, there are countless interfaces and dependencies among each other both in planning and in the construction process itself. A successful overall process therefore depends not least on the smooth transfer of information and work from the areas, which should be error-free and complete. Interfaces represent major cost drivers in the construction process because numerous pieces of information are not passed on at the required quality at the defined time. This often happens because the expectations and requirements of the respective upstream process are not clearly defined, which is why the transfer of work or information is then incomplete and faulty. This in turn leads to waste such as rework due to errors or waiting times.

Consequently, the potential in timber construction can only be fully exploited if the possibilities of the respective production are taken into account in the planning and the construction is designed accordingly. Crucial for optimal coordination from the design phase are professionals in architecture and specialist planning who are familiar with the processes of prefabrication in timber construction and have the necessary competencies to establish the manufacturing

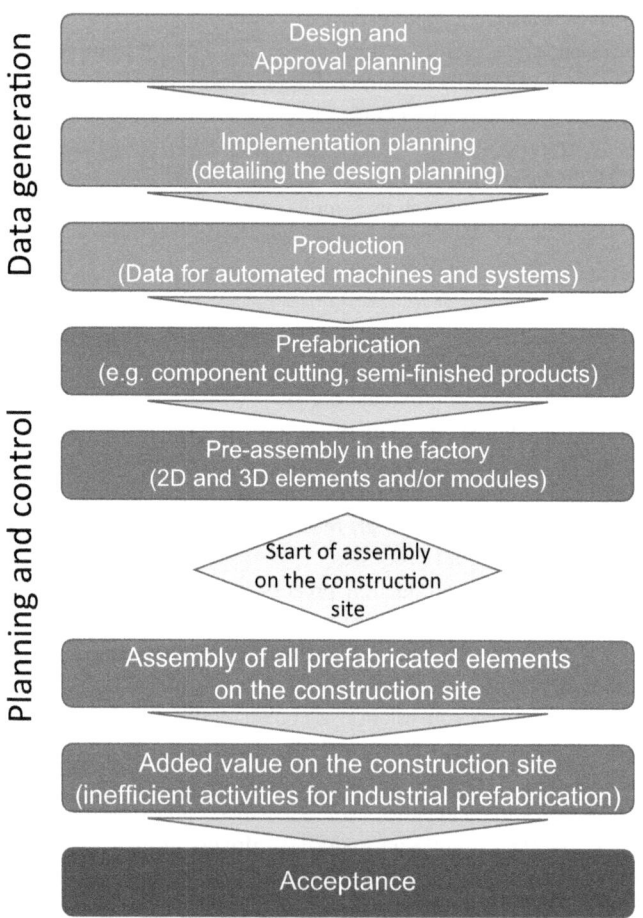

Fig. 5.1 Data process in timber construction

reference in planning. After all, the design planning has the greatest influence on all further costs of the entire prefabrication and assembly on the construction site (Fig. 5.1).

A working method that is intended to facilitate the collaboration of various departments and trades and thus improve the interfaces is Building Information

Modeling (BIM). In this collaborative methodology, a digital model of the building is created, which is used by all participants in a shared data environment throughout its entire life cycle (Messmer and Austen 2020, p. 6). So far, however, BIM is sparsely spread in timber construction, as the planners lack the necessary expertise for its use. As a result, the application usually has a negative impact on time and thus costs, instead of saving them, as is actually the goal. In addition, although a BIM model includes all information about a building, there is no reference to the actual production and no integration of production data is possible. In contrast to solid construction, the higher complexity of the constructions in timber construction also proves to be challenging, as this significantly increases the planning effort.

Furthermore, the essential service phases according to the German fee schedule for architects and engineers (HOAI) are not aligned with serial and industrial construction at the highest degree of prefabrication, as, for example, a production-oriented execution planning cannot be carried out by architects due to a lack of knowledge about detailed prefabrication.

The following will discuss the individual areas of the construction process and outline both the current approach and visions of process optimization.

5.1 Data Generation for Design and Approval Planning

This initial phase forms the basis for planning, defining how efficiently and at what cost the object can be manufactured. In order to make the most reliable statement possible, the individual possibilities of production must be taken into account in the planning. However, as it stands today, planners are often not extensively informed about the specific production conditions under which the building is industrially prefabricated. Thus, they are not aware of the effects that different designs have on the construction and assembly. As a result, production-centric planning is currently only possible to a limited extent.

Furthermore, to date, planners rarely receive manual and certainly not automated feedback on the feasibility of the designed architecture and constructions. The flow of information usually only goes from planning to manufacturing and execution, not in the opposite direction. Guidance would be provided by standard details based on manufacturing, which are made available to planners in digital form and allow them the necessary insight into prefabrication. Maximum efficiency could also be achieved through parametric planning, which is based on the requirements of the respective manufacturing and thus ensures the feasibility of implementation. Ideally, design plans would be checked by an algorithm that assesses the feasibility of the construction and also indicates the cost implications of individual planning and execution details. In this way, plans could be adjusted so that processes of prefabrication and assembly, as well as other activities in the entire construction process, become more labor and thus cost-efficient. In the approval planning, all proofs must also be provided, especially in the areas of building physics, statics, thermal and fire protection. These are strongly dependent on the chosen construction site and the planned building class and thus differ from project to project. For this reason, and because there is no sufficient standardization or sample executions, the tests are carried out individually for each project.

5.2   Data Generation for Execution Planning

The specialist planning required for the building permit (e.g., building physics), which was created in the previous phase, sets requirements for the materials to be used and the layer structure of the elements. Based on these specifications, the construction is now created with the execution planning, which includes all assembly details as well as component and material information. To take into account all production-specific possibilities, this planning is ideally carried out by the executing timber construction company.

So far, the building-specific construction details are often created individually for the production of a building, without being able to rely broadly on standard details or repeating components. As a result, different people in the planning department solve tasks or construction details in different ways. This is a time-consuming process each time and requires a high degree of expertise. In addition, all the building services must now also be planned in detail in this phase. To optimize this process, automated planning based on parameters would also be ideal here. In conjunction with machine learning, basic data could be continuously enriched, achieving increasing accuracy of automatic planning for further buildings. Using a similarity search, it would be possible to compare new details with those already created and translate them into execution plans that would only need minor adjustments if necessary. The more detailed the execution planning, the fewer execution details need to be coordinated and adjusted later in production or assembly. Ideally, all data for automated and manual processes of production and assembly are generated in this phase.

5.3 Data Generation for Production

An increasing degree of automation in production processes results in higher requirements for the quality of data preparation and provision. The construction details defined in the execution planning must now be translated into data for the machines and systems in production. Here too, efficiency could be increased by automating planning based on parameters and rules.

Currently, machines and systems often require several different data formats. However, the BTL or BTLx format is gradually becoming a standard that is being processed by more and more machine suppliers. This data standard is supported by every common software for timber construction. BTLx data contains all relevant component and processing information in detail and can thus be processed by systems in different production areas. The information required for the production processes is primarily the description of the required component processing and their installation sequence in pre-assembly. The defined assembly sequence of the components into the individual elements determines not only the manufacturing processes themselves but also the sequence of the upstream areas of material cutting, part sorting, and logistics for material provision. The overarching goal is always to have all components available in the correct form at the right time at the right workstation.

5.4 Planning and Control of Prefabrication and Pre-assembly in the Factory

A smooth manufacturing flow requires flawless production planning. Since timber construction processes are scheduled backwards, the assembly sequence of the elements on the construction site determines the entire manufacturing process. The assembly sequence dictates the logistics and thus the loading sequence of the elements, which in turn specifies which elements must be completed at which point in time. If one considers the production chain solely according to the required sequence at the end of production, it seems ideal to manufacture the elements in exactly this sequence. However, this is not feasible under real production conditions, as the manufacturing times of the individual elements vary, which particularly results in different cycle times in line production with linked work areas. For a smooth flow, the production of elements with similar manufacturing times must be grouped together (cf. line production, Fig. 3.4), which usually makes production in loading sequence impossible. A large variance in the manufacturing times of individual elements due to different complexities also makes clear timing difficult. To counteract this, there are different options for production organization. For example, the production sequence can be created with ascending and descending manufacturing times (Fig. 5.2). This measure does not allow for a uniform cycle time throughout the process, but the processing times of the elements at the respective adjacent stations are similar. This avoids longer waiting times during element handover, resulting in a smoother manufacturing flow. In contrast, there is the approach of decoupling individual assembly tables from each other through larger buffer areas. Waiting times caused by different manufacturing times of upstream or downstream processes are balanced by buffering elements in or out. However, buffer areas require a lot of space.

A Manufacturing Execution System (MES) is necessary for determining cycle times, the optimal production sequence, and the capacity planning of the different processes, which is a production control system for controlling and monitoring the processes in real time.

It provides all relevant data and information to the workstations, machines and systems, records manual or automated feedback on the progress of manufacturing, and provides a continuous overview of the manufacturing flow through data processing. With this system, it is thus possible to record every component or element and its status locally and in time. If disturbances are reported via this

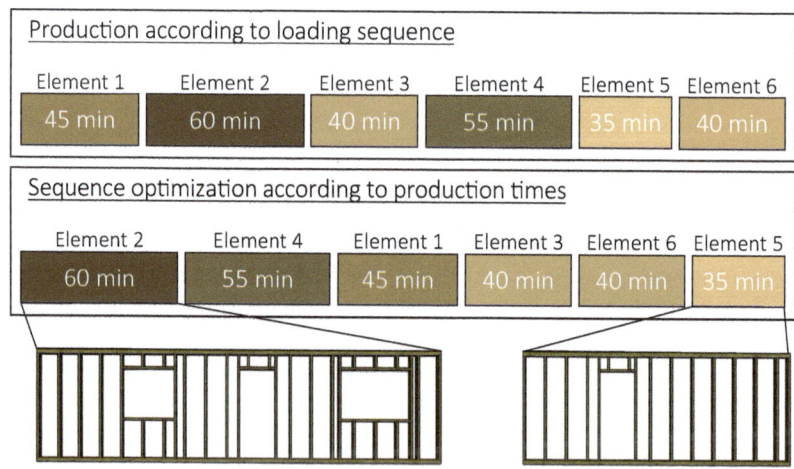

Fig. 5.2 Optimized design of the production sequence according to manufacturing times

system, the process control intervenes in such a way that the timely loading and delivery of all elements for on-site assembly continue to be ensured.

5.5 Planning and Control of Assembly and Construction Site Activities

The planning and control of assembly and construction site activities prove to be one of the most complex tasks of the overall process in timber construction. Obstacles and cost drivers associated with work on the construction site include, but are not limited to:

- Travel costs including the non-value-adding travel time of employees (especially for unplanned trips)
- Expenses and costs for overnight stays
- Reduced ergonomics
- Suboptimal workplace design
- Dependence on weather conditions
- Only limited quality control possible
- Safety

These expenses can be avoided by increasing the degree of prefabrication in the factory, which automatically reduces the work on the construction site and thus the number of interfaces to downstream work areas. The desired reduction of value-adding activities on the construction site can be achieved, for example, by using 3D modules such as complete room cells. However, this is not always possible and desired, as modules limit the architecture and make optimal use of the building plot more difficult. A noticeable trend is the combination of 3D and conventional 2D elements. 3D modules are often integrated in the form of bathroom and technical modules, which are already equipped with domestic technical devices such as heating and ventilation systems in addition to a completely finished bathroom. Although the final assembly is often only sensible on the construction site, activities from other areas can also be sensibly prepared in prefabrications, such as an automated cut of ready-made screed elements with sequential provision for laying.

Conventional construction site activities usually harbor great uncertainties in the process, as there is a lack of basic comprehensive planning. Just as there is no clear planning for the construction site, there is also a lack of feedback from the construction site to the prefabrication and planning. To optimize processes sustainably and create comprehensive transparency between prefabrication/planning and construction site, digital support similar to the MES system in production is sensible. However, this has not yet been used in depth. It is also important to mention the following points:

- The limited availability of unloading and staging areas for the prefabricated 2D or 3D Elements
  - Sometimes installing directly off the truck is impossible and a staging area is required
- Machinery and Equipment costs—a construction site usually has other machinery and equipment that take up space and require circulation areas (such as forklifts and cranes). And manuevering prefabricated elements needs to be very well synchronized with the general contractor and other trades.

CIP processes can be sensibly digitally supported. A digital workflow (e.g., the structured reporting of an error using an app on the smartphone to the work preparation) is already initiated by employees (internally or externally) on the construction site and digitally forwarded to areas of the company. This ensures that relevant information is not lost and patterns of recurring errors and problems are recognized. Only when the construction site, for example through the use of mobile devices and individually adapted apps for the support of construction site processes, becomes a fixed part of the IT processes in the company, can processes also be stabilized and continuously optimized.

Summary and Outlook 6

The trend of increasing sales figures and the ongoing shortage of skilled workers suggest no imminent change in the industry. Thus, the topics of automation and digitization for long-term optimization and support of all areas of the overall process chain will not lose their importance. As outlined above, many common, proven technologies and approaches for further solution concepts already exist. However, a sensible, efficient use of automation can only be achieved through an individual interpretation of the degree of automation in a corresponding manufacturing environment. The decisive factors for this are the specific production requirements, which are particularly influenced by factors such as the required performance, the level of investment, and the possible use of personnel. Furthermore, the possibilities for integration into existing structures, the future viability, and the flexibility of the systems play a crucial role in the selection of technology.

The requirements for digital processes increase with a higher degree of automation. The error-free creation of construction plans and details, as well as extensive data generation and processing, are becoming increasingly important. Only in this way can waste-free processes in prefabrication be guaranteed. However, the described developments in the timber construction industry not only affect timber production, but also the entire planning and construction process. The described difficulties at the interfaces between different specialist areas and trades can only be overcome through careful, comprehensive information exchange. The digital collection and provision of real-time data, which enable transparent communication, are predestined for this. A prerequisite for fully exploiting the potential in the processes is the collaboration of all areas of responsibility from design planning to building acceptance, the operation of the building, and even building maintenance.

A. Heinzmann and N. Karatza, *Automation and Digitization in Timber Construction*, https://doi.org/10.1007/978-3-658-47130-9_6

The challenge with increased use of automation and digitization is the implementation of highly individualized, complex buildings. Here, there is still a high proportion of manual interventions. The solution is seen in the concept of mass customization, which describes the mass production of individual products. The implementation in timber construction begins with the standardization of basic processes and executions. This includes both planning and organizational activities as well as assembly techniques. While decisive, non-visible components for customers such as element structures and support systems should be standardized, there is a need for freedom in the area of external appearance such as facades and floor plan design based on defined guidelines. In this way, manufacturing processes can be uniformly automated, allowing for industrial element production with individual design possibilities.

In summary, it can be said that in timber construction, both the market development and the increasing possibilities of automation and digitization solutions are seen as having enormous potential.

Quellen

A

Bauernhansl, T. (2020). *Fabrikbetriebslehre 1: Management in der Produktion* (1. Aufl.). Springer Vieweg.

Dolezalek, C. M. & Baur, K. (1973). *Planung von Fabrikanlagen*. Springer Publishing.

Eversheim, W. & Schuh, G. (1999). *Produktion und Management*. Springer Publishing.

Holzbau Deutschland. (2022, Mai). *Lagebericht 2022*. Holzbau Deutschland – Bund Deutscher Zimmermeister im Zentralverband des Deutschen Baugewerbes e. V. https://www.holzbau-deutschland.de/fileadmin/user_upload/eingebundene_Downloads/Lagebericht_2022.pdf

Jakob, S. (2021, 6. Juli). *Hohe Nachfrage: Deutschland geht das Holz aus.* tagesschau.de. https://www.tagesschau.de/wirtschaft/technologie/holz-baustoff-mangel-corona-101.html

Karatza, N. P. (2019, August). Development of a compact flexible manufacturing cell for producing timber frame elements. *IWMS-24 Proceedings*, 209–216.

Messmer, B. & Austen, G. (2020). *BIM – Ein Praxisleitfaden für Geodäten und Ingenieure: Grundwissen für Geodäten und Ingenieure (essentials)* (1. Aufl.). Springer Vieweg.

Ohno, T., Rother, M., Stotko, E. & Hof, W. (2013). *Das Toyota-Produktionssystem: Das Standardwerk zur Lean Production* (3. Aufl.). Campus Verlag.

Schankula, A. (2012). Vorgefertigtes Bauen mit Holz. *Detail, 6/2012*, 622–669.

Statistisches Bundesamt. (2020, Juli). *Bauen und Wohnen, Baufertigstellungen von Wohn- und Nichtwohngebäuden (Neubau) nach überwiegend verwendetem Baustoff, Lange Reihen ab 2000.* https://www.destatis.de/DE/Themen/Branchen-Unternehmen/Bauen/Publikationen/Downloads-Bautaetigkeit/baufertigstellungen-baustoff-pdf-5311202.pdf

VDI 2510 Blatt 1: 2009–12 – Infrastruktur und periphere Einrichtungen für Fahrerlose Transportsysteme (FTS)

VDI 3415 Blatt 1: 2018–10 – Entwurf – Holzbearbeitungsmaschinen – Prozessqualifikation Maschinenabnahme

Batch number: 08422745

Printed by Printforce, the Netherlands